Practicing Medicine Without A License ?

The Story of the Linus Pauling Therapy for Heart Disease

The Pauling Therapy Handbook

Volume I

ISBN: 978-1-4357-1293-5
Published by Lulu.com

No one other than Owen Fonorow is responsible for the
contents of this book. The statements in this book that are not
direct quotes are entirely his opinion. He is not suggesting that the
reader do anything, but rather he is suggesting that should the
reader decide to follow others down the Pauling path, he should do
so with his eyes open and at his own risk.

The author pledges to donate a substantial percentage of the
royalties paid on the sale of this book to The Vitamin C
Foundation, a Texas nonprofit corporation designated as a qualified
501(c)(3) corporation by the Internal Revenue Service.
http://www.vitamincfoundation.org.

The VITAMIN

FOUNDATION

Vitamin C Foundation Books

Practicing Medicine Without A License ?

The Story of the Linus Pauling Therapy for Heart Disease

The Pauling Therapy Handbook

Volume I

Owen Fonorow

with

Sally Snyder Jewell

"Nothing, not all the armies of the world, can stop an idea whose time has come." — Victor Hugo

This book is dedicated to the millions who have perished before their time.

Milton, we miss you.

Table of Contents

A Tribute to Linus Pauling

While enjoying a delicious breakfast of bacon and eggs with my son one morning I mentioned my difficulty in writing this tribute to the man whose books are probably the reason I am still alive. I explained that several books would be required to say everything that should be said about Linus Carl Pauling, the world's only two-time unshared Nobel Prize recipient and one of its greater scientists, and that entire books had already been written. My son responded, "Your book is the tribute." So this will be brief.

To Linus Pauling: Thank you for freeing me of all prescriptions and over-the-counter drugs and for making the past 20 years so unlike the first 30.

To the Linus Pauling Institute: May you see the light once again.

To the National Academy of Sciences: Act like scientists.

To the medical scientists in colleges and universities: Run a study. People are dying.

To the heart patients: Think for yourselves.

Owen Fonorow

Introduction

State permission should not be required in order for a person to offer nutritional advice to a friend, relative, or acquaintance. This book, *Practicing Medicine Without a License? The Story of the Linus Pauling Therapy for Heart Disease*, is more than the story of Linus Pauling's invention of a nutritional treatment for cardiovascular disease. Owen Fonorow poses many thought-provoking questions that beg answers.

For example, why have medical scientists ignored Linus Pauling's heart disease claims, and why, after more than a decade, are there still no studies designed specifically to test the vitamin C theory of heart disease? Why hasn't the apparently curative Pauling therapy protocol been thoroughly investigated?

What if someone recommends vitamin C to a heart patient and the patient is cured? Was that advice lawful? What about when a person then tells a third person what occurred in the first case? When is a person who recommends nutritional advice acting lawfully or unlawfully?

The Food and Drug Administration (FDA) rules have been shrewdly designed to ban nutritional supplements that have not been approved for use as drugs. The current law can be construed as making us all criminals if we give nutritional advice, especially if we recommend supplements that cure specific diseases. Even licensed medical doctors cannot recommend "unapproved" nutritional supplements. This interpretation is derived from the incomprehensible rules promulgated by the Food, Drug, and Cosmetic Act, which defines a "disease" and a "drug."

According to the Food, Drug, and Cosmetic Act, a disease is vaguely defined as any unusual condition of the body or mind. The dictionary definition of disease includes pain, obesity, and muscular weakness. A "drug" is defined as any article "intended for use in the diagnosis, cure, mitigation, treatment or prevention of disease."

All nutritional supplements, called nutriceuticals, are in most instances intended to prevent, cure, or treat a disease. All vitamins are intended to prevent and cure their deficiency disease, as are all minerals, essential amino acids, and so forth. The same goes for probiotics and herbs. They are all intended to treat disease, and therefore they might all be misconstrued as drugs. All drugs that have not been approved by the FDA are banned.

The 1994 Dietary Supplement Health Education Act (DSHEA) of Congress forced the FDA to exempt nutriceuticals from its approval requirement, clearing the way for their lawful manufacture. However, the law has not clarified what can legally be said about what these lawful products do. The murky law paints the picture of being arrested, hand-cuffed, and taken off for prosecution should someone make any statement resembling a health claim regarding an "unapproved" nutriceutical.

If all nutritional products might be construed as drugs, should not the law be clarified when so many are being made to appear as criminals by recommending their use, including medical doctors when they prescribe some unapproved treatment, naturopathic physicians, acupuncturists, homeopaths, herbalists, chiropractors, friends telling friends, relatives telling relatives, and mothers advising their children?

Are employees of vitamin stores practicing medicine without a license when they recommend nutritional products? Clearly, the answer is they are not. Certainly, such activity cannot be considered practicing medicine without a license because medical licenses are issued by States, not by the Federal Government. A person can only be guilty of practicing medicine without a license if the act, in this case recommending a nutriceutical, is defined as practicing medicine under the State's license law. States may prohibit unlicensed individuals from recommending drugs; however, it is doubtful that States include in their license laws prohibition of the use, sales, manufacture, administration, or distribution of nutritional products.

As manufacturers are free to offer nutriceutical products under the current FDA rules, what rule or law restricts an unlicensed person from recommending a nutriceutical? Most magazines and radio stations carry articles or ads that recommend nutritional products. Almost every infomercial in the world recommends a "banned" drug (nutriceutical) that will cure you from a disease, stop your pain, make you lose weight, or build up your muscles. Thankfully, the answer is that recommendations of this sort are statements that should be protected by the Freedom of Speech provision of the First Amendment to the United States Constitution.

The people of the United States of America should not fear prosecution as a result of recommending nutritional supplements that happen to cure disease. The current laws do not yet take into account that serious diseases may now be cured by obvious non-drugs. It is time for laws to be enacted that reflect the new realities presented in this book — laws that define what legally constitutes a drug, not on the basis of its curative effect but on the basis of the danger posed by the ingestion of the substance.

William Decker, President
Tower Orthomolecular Laboratories
Las Vegas, Nevada

Preface

"I solemnly profess that I hate all pretenses to secrets and I look upon the printed bills of quacks, who pretend to nostrums and private medicines, to be mere cheats and tricks to amuse the common people and to pick their pockets. But if any man can communicate a good medicine, he shows himself a lover of his country more than of himself, and deserves the thanks of mankind." — Dr. William Simpson (1680 AD), found in Ralph Moss's *Cancer Therapy: The Independent Consumer's Guide*

This book provides important information for those who have suffered a heart attack or stroke; undergone coronary artery bypass surgery; and/or have high blood pressure or any other form of cardiovascular disease (CVD). Those in this category number in the millions. According to some estimates, nearly half of the adult population of the United States has some form of the disease.

For the past 12 years persons suffering from heart disease have called and asked for advice. These people typically recover within 30 days and the majority experience significant relief within as little as 10 days.

I am not a medical doctor. In fact, this book could not have been written by one. Licensed medical doctors are prohibited from dispensing drugs and treatments that have not been approved by the Food and Drug Administration (FDA). Furthermore, your doctor is not about to recommend, and the FDA is not about to approve, a treatment with dosages higher than the upper tolerable limit as established by the U. S. Government. The upper tolerable limit of vitamin C has been established at 2,000 mg per day, which is arbitrary and well below what Linus Pauling and others have recommended. Results such as those reported in Chapter 8 of this book will not be seen if one limits his vitamin C intake to the upper tolerable limit.

My field is computer science. I worked for and retired from AT&T Bell Laboratories, having joined Bell after graduating from the United States Air Force Academy and completing my service. In 1994 I realized that Linus Pauling had made an important discovery that was being ignored. This discovery was of the utmost importance because Pauling advocated a non-toxic, non-prescription treatment for America's leading cause of death. It isn't often that a world-renowned chemist makes the claim that "cardiovascular disease can be controlled, even cured."

The reason that Linus Pauling is ignored by the medical profession will become obvious: *A doctor isn't required to administer the treatment.* This book contains world-changing news. The economic impact over time amounts to trillions of dollars, on par with the National Debt. To date, the mainstream media has ignored this story.

We ask you to suspend disbelief for a moment because what follows is even harder to swallow. In our opinion, after the first few chapters of this book the reader will have a better understanding of heart disease than the average heart specialist who has trained for more than ten years. He will understand what really causes heart disease (cardiologists do not); he will know how to prevent heart attacks and strokes (cardiologists do not); and he will know the basic elements of how to treat the disease with non-toxic, non-prescription nutrients (cardiologists will not).

Sadly, Linus Pauling died in 1994 at the age of 93 years. I picked up the ball when the Linus Pauling Institute of Science and Medicine let it drop. I do not mean to diminish the contributions of many others. Linus Pauling's close associate, Matthias Rath, M.D., was the first Director of Cardiovascular Research at the Linus Pauling Institute of Science and Medicine. Dr. Rath has been a significant force behind the promotion of the vitamin C theory of heart disease, and his companies and Institute are spreading the message to millions of people around the world.

However, in my opinion, Rath and other early advocates seem to be missing perhaps Pauling's most important point: *'It's the dosage, stupid.'* The dosages of vitamin C and lysine that Linus Pauling recommended were far higher than what any other advocate was then recommending. I coined the term "Pauling therapy" to represent the therapy administered at the recommended high dosages and to honor its inventor.

People began to call and ask for advice after we posted preliminary news of the Pauling discovery on the Internet. Wary, we told them to watch Linus Pauling on a video lecture he had recorded and pay special attention to his recommended dosages. We also created a physician referral list so that we could recommend knowledgeable doctors to people with specific medical questions. The difficult part was finding such doctors.

In the beginning, the only people who called were in pain and seemingly without recourse. Their doctors, unable to control their pain, had advised these callers to search the Internet for possible alternatives. We quickly discovered that nearly all heart patients who adopted Pauling's specific recommendations recovered, and quickly.

As we grew more confident in what Pauling had recommended, I began to write a series of articles about the Pauling/Rath discoveries, and many were published in an obscure alternative medical journal, *The Townsend Letter for Doctors and Patients*. This book is a compilation of more than 30 papers written for journals of alternative medicine. The story has been picked up by some of the newer alternative media.

William Decker is an entrepreneur who was also a personal friend of Linus Pauling. Mr. Decker learned about the existence of the Pauling video lecture from *The Washington Times*. A few years earlier Pauling had informed Decker of the experiments then being conducted at the Pauling Institute, stating, *"If the outcome of Lp(a) experiments turns out as we expect, the entire problem of cardiovascular disease can be eliminated."* These words uttered by Linus Pauling *"burned the term Lp(a) into my brain,"* Decker later told us. In 1996 Decker formed Tower Laboratories Corporation, which became the first company to offer "Pauling therapy" products. No other company was willing to match Pauling's recommendations, which generally exceeded the U. S. Government-inspired "maximum tolerable limit." Many of the case studies in Chapter 8 were with this Pauling therapy formula, Tower's *Heart Technology*™.

We know quite a bit about the Pauling therapy, but there is much more to learn. There are thousands of studies on vitamin C but not at the dosages recommended by Pauling or in combination with lysine. Until now, few medical researchers have been willing to participate in such. We hope this book will help change that mindset.

We also offer our innovative solution to universal health care that would reorient free market forces to reward health and not illness. The *Vouchers* proposal presented in Chapter 12 would save the United States billions of dollars in health care costs. Such a plan is feasible because of low-cost, natural cures such as the Pauling therapy.

This story also has a dark side. Over the years we have come to realize that someone or some group has deliberately disseminated false medical information in the media disguised as news stories. These stories were designed to discourage the use of vitamin C supplements. It is hard to understand how persons engaged in this deception could rationalize their behavior.

As we attempted to make this news public, it slowly dawned on us what Pauling's discoveries were up against. I unwittingly followed in the footsteps of many alternative doctors. I finally won the legal battle in an Illinois courtroom with the assistance of health consumer advocate Tim Bolen. The story of *The Great Suppression* is intertwined with government interference (the FDA) and failure (the NIH) and is the subject of The Pauling Therapy Handbook, Volume II.

There are many people in Federal custody for selling unapproved cures for major diseases, especially cancer. Curing people is more than politically incorrect; doing so without a license is contrary to many laws in the United States and can result in incarceration, no matter that the extremely ill are brought back from the brink of death as a result.

"I just began doing the Pauling program for heart disease and had a tremendous relief in symptoms in 10 days." — Tanya Bartunek, Sunrise Farm, 2004

Chapter 1

The Problem With Cardiology

The medical specialist who treats heart patients is a cardiologist. Typically, a cardiologist must attend four years of medical school, three years of residency in internal medicine, and three years of fellowship training in cardiology. According to statistics published by the American Heart Association (AHA), these specialists are involved with the care of more than 50 million American adults who have been diagnosed with some form of heart disease.

American cardiologists use a broad range of diagnostic procedures and prescribe from among a myriad of heart medications that have been approved by the FDA. Well-equipped cardiologists are competent to diagnose heart disease from blood tests. According to news reports, nearly half of emergency room visits for chest pain are not for heart attacks. These common pains are usually from indigestion.

The best selling prescription drugs are the cholesterol-lowering *statin* drugs prescribed by cardiologists such as Lipitor®, Zocor®, and Crestor®. These drugs are thought to be so beneficial that doctors will even prescribe them for people with low cholesterol. The pharmaceutical company Merck & Co., Inc. has petitioned the FDA to allow its statin drug Mevacor® to be sold over the counter without a prescription.

Collectively, the statistics published by the AHA and the Center for Disease Control (CDC) are appalling, in that cardiologists and cardiac surgeons together performed more than 900,000 heart operations in the United States during 1996, i.e. 500,000 coronary artery bypass graft procedures and 400,000 balloon angioplasty procedures. Symptoms of the disease reappear in nearly half of the heart patients who undergo these types of procedures. Therefore, the treatment was a failure in nearly 360,000 patients.

Restenosis is the term used by cardiologists for the regrowth of atherosclerotic plaques after heart surgeries. In vain attempts to stop restenosis cardiologists often resort to the routine insertion of hollow metal scaffolds called bare metal stents into the coronary arteries and bypass grafts. However, stents have not solved the problem of rapid reocclusion.

The problem, in our opinion, is that cardiologists do not understand the nature of the disease they are licensed to treat. If they did understand, so many would not routinely advise their patients against vitamin C.

This year alone, heart disease and stroke will kill more than 700,000 Americans.

Why Are Cardiologists Anti-Vitamin C?

Pauling therapy advocates have, over the past 14 years, received hundreds of reports from heart patients who have self-administered the Pauling therapy. Richard's is one of the more recent cases and you can read his full story in Chapter 8:

My medical history ranks as at the extreme end of desperate. I started using the Pauling therapy about four months ago. I gradually added more of the elements of the protocol as I felt better. I am now on the full protocol as described by Linus Pauling, et al. I have not felt this good in over twelve years. My first heart attack was '95, the last was '03. I have had five operations and my doctors have never been able to explain what was killing me. I had tried everything including a strict vegan diet, etc., for five years straight and with no cheating. Nothing worked, I just got worse. Now I know that vitamin C was the missing piece of the puzzle. I also know I will probably live a healthy life for several more years free of artery disease. Can you imagine how I feel? — **Richard, March 2006**

The Pauling/Rath vitamin C theory and its associated treatment is a profound economic threat to pharmaceutical special interests, and this may be the reason that this story has never received publicity. Another reason for this neglect may be that it sounds too good to be true, regardless of the fact that a famous Nobel Prize recipient and chemist invented this non-prescription, non-toxic alternative and there would be potentially enormous public health benefits. Most of us believe that if true, a development of this magnitude would never be ignored.

A Thought Experiment - What is the Real Cause of Heart Disease?

I teach computer science to college students. Computer-oriented material is dry, so from time to time I change the pace and ask my students a deceptively simple question: "What do you think causes heart disease?"

Pause and ask yourself the same question. What is the cause of the narrowing and hardening of the arteries that eventually leads to high blood pressure, heart attack, and stroke?

The students brainstorm about possible reasons and their answers usually include the following:

Stress
Poor diet
Fatty foods
Cholesterol
Lack of exercise
Too much exercise
Junk foods
Aging

Next we discuss the United States death rates due to heart disease from 1900 to 1996 (see Figure 1). I draw the graph on the blackboard. The cardiovascular disease mortality rate peaked between approximately 1950 and 1960, and then around 1970 the death rate from cardiovascular disease turned around and declined by 30 to 40 percent.

The students are then asked to explain the steady increase in the heart disease mortality rate from the years 1900 to 1950 followed by a decline, keeping in mind their previous answers as to the cause of heart disease. Whatever the reason, the United States was the only industrialized country to see a decline in its heart disease mortality rate. In terms of raw numbers, almost 300,000 fewer died in 1996 than in 1970, even though the population nearly doubled. Roughly 700,000 people died in 1970, while the Nation's health authorities estimate that only 400,000 died from cardiovascular disease in 1996.

This begs the question, "What happened in 1970?" Again the students brainstorm. There are not many good answers to the question of why the United States was the only industrialized nation to experience such a dramatic reduction in its death rate.[1]

FIGURE 1. Age-adjusted death rates* for total cardiovascular disease, diseases of the heart, coronary heart disease, and stroke,† by year — United States, 1900–1996

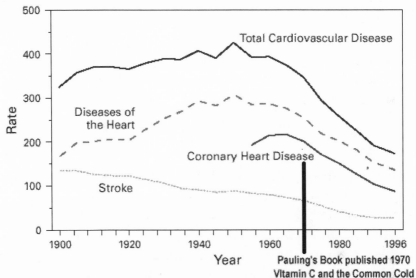

*Per 100,000 population, standardized to the 1940 U.S. population.
†Diseases are classified according to *International Classification of Diseases* (ICD) codes in use when the deaths were reported. ICD classification revisions occurred in 1910, 1921, 1930, 1939, 1949, 1958, 1968, and 1979. Death rates before 1933 do not include all states. Comparability ratios were applied to rates for 1970 and 1975.

Source: Adapted from reference 1; data provided by the National Heart, Lung and Blood Institute, National Institutes of Health.

The students generally offer rather weak ideas such as the following:

---

[1] All material in the Morbidity and Mortality Weekly Report (*MMWR*) series is in the public domain and may be used and reprinted without special permission; citation as to source.

McDonald's hamburgers
Man landing on the moon
The UNIX operating system
The end of the war in Vietnam
Advances in medicine
Medical technology
Prescription drugs (Note: Cholesterol-lowering medications first appeared in the 1980s, and the mortality rate has been creeping up since that time.)

The dramatic drop in the CVD mortality rate in the United States is a historical fact. It is not easy to account for, which may explain why we rarely hear about this in the media.

The Linus Pauling Institute of Science and Medicine (LPI) has suggested that Linus Pauling himself may have had something to do with the decline. The Institute notes that this reduction in the mortality rate only occurred in the United States after Pauling wrote his book, *Vitamin C and the Common Cold*. This book was first published in 1970 and it soon became a National best seller. Furthermore, LPI reports that the consumption of vitamin C increased by 300 percent in the United States during the 1970s after publication of the book.

The Pauling book on vitamin C and the subsequent increase in vitamin C consumption may not explain the phenomenon, or it may not be the only explanation, though any other explanation has to be able to account for the 40-percent decline in the mortality rate and why it occurred in America but not in other industrialized countries.

In 1992 an important study was published which confirmed that this effect was probably related to vitamin C. Dr. James E. Enstrom and colleagues studied a large quantity of food intake data (from over 11,000 people) that had been collected by public health authorities. The data included information on vitamin C consumption, but for unknown reasons this part of the data had never been analyzed. Enstrom determined that increased intake of vitamin C had reduced the death rate from heart disease by nearly half and prolonged life for more than six years.

At this point in my discussion with the students we review ground-breaking experiments that were conducted more than 50 years ago in Canada. The Canadian research group, led by George C. Willis, M.D., was the first to study the relationship between vitamin C deficiency and heart disease. Dr. Willis was the first to blame atherosclerosis on the mechanical stress caused by the beating heart. Those experiments are the subject of the next chapter. During the ten-minute lecture in which I summarize the Canadian experiments my students learn more about vitamin C and heart disease than the average cardiologist apparently learns in ten years of advanced schooling.

A prerequisite to understanding the significance of the Willis experiments is the knowledge provided by Linus Pauling, knowledge that is evidently not taught in medical schools. Over the course of the history of life on earth, few species have lost their ability to make vitamin C and survived. Today, practically all animals have the ability to make vitamin C in their bodies. Dogs, cats, cows, and birds all make vitamin C. Humans cannot. We share this rare inability with only a handful of other species including the guinea pig and a number of high order primates.

Humans and guinea pigs must obtain vitamin C in the diet to exist. Therefore, the guinea pig is an ideal subject for the study of the vitamin C deficiency disease scurvy (and, as it turns out, atherosclerosis). The species that can make their own vitamin C do so throughout the day. Most animals create the vitamin in the liver and it is injected directly into the bloodstream.

The Canadians had been the first to observe that arterial plaques (atherosclerosis) growing in and clogging human arteries are not randomly distributed throughout the bloodstream. Over and over plaques appear uniformly in the same locations — inside the walls of arteries that are closest to the heart. The Canadians were aware that collagen is required to keep arteries strong and that vitamin C is required for adequate collagen synthesis. I ask my students to think about the heartbeat and its effect on the coronary arteries as it squeezes and releases over and over, on average between 60 and 100 times per minute in an adult.

Doctor Willis reasoned in one of his papers that the physical force exerted on the walls of arteries by the heartbeat provides the best explanation; he theorized that the plaques containing cholesterol and calcium are the body's response to the wear and tear — a natural healing process whereby the body shores up arteries that are beginning to weaken.

The Willis team ran a series of brilliant experiments to validate his hypothesis that a vitamin C deficiency is the root cause of the disease process. Their first guinea pig experiment was repeated by Linus Pauling and his associate, Matthias Rath, M.D., in 1989.

If cardiologists or medical journal editors are aware of the connection between vitamin C and the strength and elasticity of arteries, they keep this information to themselves.

For my students, the *coup de grâce* is learning that vitamin C has no known lethal dosage level. A lethal dosage level of 50, for example, is the dosage of a substance that kills 50 percent of the animals given that dose. A laboratory animal cannot be killed with any dosage of vitamin C (and for that matter, there is no known lethal dosage level for the amino acid lysine, which is discussed in Chapter 6). It is not unusual for students to come to our next class session carrying a bottle of vitamin C!

Strangely, the Canadian work just summarized and the Pauling/Rath vitamin C *Unified Theory* are entirely foreign to the average cardiologist. Your cardiologist was never exposed to any of this information during his years of schooling or in his medical journals. We infer that few medical textbook writers are aware of the Willis work. If they are, they would probably make the excuse that the study in humans was small or that other studies with low dosages showed no improvements. However, there are few if any other studies that directly measure vitamin C intake with the progression or regression of atherosclerosis.

The reasons and methods for barring the Pauling/Rath science from major medical journals are discussed in The Pauling Therapy Handbook, Volume II.

Judging from what patients relate to us about what their cardiologists have told them, we infer that cardiologists are taught that reversal of atherosclerosis (hardening of the arteries) is impossible. These doctors may have even been taught that vitamin C can be harmful to heart patients.

Cardiologists Can Learn

We often wonder why cardiologists ignore Pauling's research. The following incident occurred in 1996 or early 1997 and may provide a clue.

A pharmacist, let's call him Dan, found our Web site and called us. We discussed the Pauling theory as well as the reports up to that time from people who had adopted the therapy.

It turned out that Dan was a leading pharmacist. He made it his business to study new pharmaceuticals in the pipeline. His lectures to doctors about drug advances were being heard by packed houses. Dan had called because he was about to give a lecture to a group of cardiologists and he had browsed the Internet to keep abreast of leading-edge developments.

During our conversation Dan related his new plan to tear up his normal lecture and instead give the cardiologists a lecture about the Pauling/Rath *Unified Theory*. When he told me this, it was a little disconcerting, though I wished him well. Dan called back after the lecture and told me the following story.

The lecture had gone well and the vitamin C theory was understandable. However, after the lecture, livid cardiologists stood to their feet. Their anger was not directed at Dan or his presentation, but rather their animosity was directed toward the medical journals. The cardiologists were angry because it was the first time they had heard about the Pauling research from a respected source. *"Why haven't we heard any of this before?"* was the common response.

I predict that before long, perhaps after you have heard tales of remarkable case histories (maybe even your own case), you will come to realize that one man can be credited for saving the lives hundreds of thousands of his fellow citizens since 1970.

Linus Pauling was an American hero.

"The principle of science, the definition, almost, is the following: The test of all knowledge is experiment. Experiment is the sole judge of scientific 'truth'." — Richard Feynman, 1965 Nobel laureate in physics

Chapter 2

Willis

George C. Willis, M.D., was a medical doctor and the leading physician in a Canadian group that published a series of papers about vitamin C and heart disease in the peer-reviewed *Canadian Medical Association Journal* during the 1950s. These landmark papers are of a series of experiments that "prove" that a single variable — low intake of vitamin C — causes the condition commonly called atherosclerosis which narrows arteries, restricts blood flow, and causes angina pain.

Ascorbic acid is the technical name for vitamin C, and the two terms are used interchangeably in this chapter and throughout this book. A *scorbutic* diet is a diet that is low in vitamin C and will cause *scurvy*, the classic vitamin C deficiency disease. *Lesions* are injuries or abnormalities in the walls of arteries that are thought to precipitate the disease process.

The pioneering research into the relationship between vitamin C and heart disease was begun in the 1940s not long after the molecular structure of vitamin C was determined, circa 1937. Before World War II the Canadian J. C. Paterson had published papers in the *Canadian Medical Association Journal* with his findings that capillaries tend to rupture and hemorrhage when they become depleted of vitamin C. Paterson was not published after the war.

The observation that weaker arteries are deficient in vitamin C was later verified by Dr. Willis, who examined postmortem tissues for the presence of the vitamin. Willis found that the levels of vitamin C in the tissues of those who had died suddenly, e.g. from a car accident, were higher than the levels of vitamin C in the tissues of those who had died slowly in hospitals. Of the latter, many were entirely depleted of vitamin C.

Willis began his experimentation with guinea pigs. These animals lack the ability to manufacture their own vitamin C, a trait they share with humans.

After the initial results were published supporting the theory that a vitamin C deficiency is the cause of atherosclerosis in guinea pigs, Willis conducted the first known tests of whether or not vitamin C deficiency causes atherosclerosis in humans. The outcome of this pilot study was remarkable for both the results and the methods used to investigate narrowing of the coronary arteries. This work, long forgotten, should have stimulated more research. We can now surmise that with higher, Pauling-like dosages of the vitamin the results would have been even more remarkable.

Finally, in 1957 Willis returned to experiments with the guinea pig. His landmark *Reversibility* experiment clearly showed that early atherosclerosis is reversible simply by increasing vitamin C to optimal levels.

Willis pointed out how the lesions in the sacrificed guinea pigs were like human lesions but unlike the lesions that can be generated in rabbits on very high cholesterol diets.

The guinea pig experiments have been repeated, most recently by Linus Pauling and Matthias Rath, M.D., in 1989.

The Pauling/Rath experimental results were published in the *Proceedings of the National Academy of Sciences.* The regular medical journals have consistently refused to publish research along these lines, though Linus Pauling had the right to have his papers published in the *Proceedings*, as do all members of the National Academy of Sciences. In this manner Pauling was able to publish around the barrier that has been erected to shield the medical profession from this work. Although some of their work was published in this prestigious publication, not every paper was accepted. It is a matter of record that the *Proceedings of the National Academy of Sciences* refused to publish the Pauling/Rath vitamin C *Unified Theory* paper.

There is now repeatable scientific evidence that a single deficiency in the diet of the guinea pig, i.e. low vitamin C, results in rapid atherosclerosis. Willis was the first to offer supporting evidence that the same process takes place in humans. This chapter reviews those early and important experiments.

When the extensive studies made on the subject of atherosclerosis have been reviewed, it is apparent that only the morphologic features form a basis common to all. These morphologic features are first of all a disturbance in the intercellular ground substance of the arterial intima at points of mechanical stress. Stainable lipids are then deposited in the altered ground substance. Macrophages appear and phagocytose lipid, and capillary sinusoids arising from the media of the arterial lumen invade the intima and may give rise to intimal hemorrhage. Finally thrombosis may occlude the artery already narrowed by these processes. — **G. C. Willis, 1953**

The Willis Mechanical Stress Theory

Dr. Willis reasoned that only the mechanical stress caused by the pulse could explain the typical pattern of atherosclerosis so often observed in heart patients. Using high school physics, he calculated the forces that govern the load on arteries. He correlated the maximum mechanical stress with sites of atherosclerosis in experimental animals, with vascular anomalies in the human, and finally with atherosclerosis in the common case.

To Willis, the body was laying down plaque precisely where needed to stabilize the vascular system. Willis argued that the forces on arterial walls are the strongest where arteries bend and where they split into two. These are the same locations near the heart where the plaques of atherosclerosis form and where weaknesses can be induced by the first lesions of the vitamin C deficiency disease scurvy. These profound observations regarding the role of mechanical force and the susceptibility of arteries to these forces when vitamin C is deficient in arterial tissue are at the heart of the Pauling/Rath vitamin C *Unified Theory*.

The role of vitamin C in promoting collagen synthesis and maintaining strong, healthy arteries was understood by the early 1950s. Willis noted that scurvy had been characterized as a disturbance of the intimal ground substance. The *intima* refers to the inner lining of a lymphatic vessel, an artery, or a vein. Willis decided to run the first experiments to test the idea that chronic scurvy was the most likely cause of the heart disease process.

An Experimental Study of the Intimal Ground Substance in Atherosclerosis (G.C. Willis, Canad.M.A.J. July 1953, Vol 69, pp 17-22)

The Willis experiment with guinea pigs was published in the *Canadian Medical Association Journal* in 1953 under the title, *An Experimental Study of the Intimal Ground Substance in Atherosclerosis.* Could heart disease be a mechanical problem exacerbated by a vitamin C deficiency? Willis decided to find out.

Guinea pigs are prone to scurvy and they easily develop occlusive cardiovascular disease. The experiment utilized 143 guinea pigs that were divided into various groups and sacrificed at varying intervals. The results are summarized in Table 1.

Table 1: Results in Each of the Dietary Groups

Diet	Number of Animals	Number of Animals With Atherosclerosis	Number of Animals Without Atherosclerosis
Fed chronic scorbutic (low vitamin C)	20	9	11
Fed chronic scorbutic with oral vitamin C (ascorbic acid)	22	0	22
Fed chronic scorbutic with oral vitamin C and cholesterol	18	16	2

Fed chronic scorbutic with IV vitamin C and cholesterol	18	7	11
Fed acute scorbutic	32	19	13
Fed acute scorbutic with oral vitamin C	16	0	16
Fed acute scorbutic with cholesterol	11	11	0
Fed acute scorbutic with oral vitamin C and cholesterol	8	4	4

According to the paper, scurvy, both acute and chronic, was effective in producing the lesions which have been described in early human atherosclerosis. These lesions developed in as few as 15 days from the onset of the scorbutic diet. From the paper, "This represents a very short time when it is recalled that it takes about 12 days to produce ascorbic acid depletion in the guinea pig. The lesions occurred at normal cholesterol levels and were not accompanied by lipid deposits in the spleen."

Willis noted in this paper how similar the atherosclerotic lesions of the deprived pig were to human lesions and how those lesions were unlike the "fatty streaks" that could be created in experimental animals that were fed ultra-high cholesterol diets. "Scurvy is an ascorbic acid deficiency disease with a resulting disturbance of the ground substance," Willis wrote. "I have shown that scurvy is effective in producing the lesions in the guinea pig arteries which are morphologically identical with human atherosclerosis."

Willis concluded that vitamin C is essential for the maintenance of the ground substance around the arterial intima. "Any factor disturbing ascorbic acid metabolism, either systemically or locally, results in ground substance injury with subsequent lipid deposits."

From this experiment we know that a diet restricted in vitamin C can be used to induce atherosclerosis of the type commonly found in humans.

Serial Arteriography in Atherosclerosis (G.C. Willis, A.W. Light, W.S. Cow, Canad.M.A.J. Dec 1954, Vol 71, pp 562-568)

In his first paper Willis had theorized that plaque buildup is the healing response to a repeated insult — the heartbeat. He ran experiments that showed vitamin C depletion in tissues and the atherosclerosis that can be induced in the guinea pig by depriving it of vitamin C.

At this point Dr. Willis was ready to run tests in humans and wished to investigate whether giving vitamin C to his heart patients would have the same effect as when given to the guinea pig. The problem was how to measure outcomes, since it isn't possible to sacrifice human patients and examine their arteries.

Necessity being the mother of invention, Willis invented a technique he called *serial arteriography* that used x-rays and a special dye to peer inside human arteries. For the first time medical doctors were able to see inside the arteries of a living being and observe the progression and regression of atherosclerotic plaques. The pictures he published were remarkable.

The 16 patients studied by arteriography were selected from the Queen Mary Veterans' and St. Anne's Hospitals in Montreal. All were men who varied in ages from 55 to 77, with the average age being 64 years, and who had shown many of the clinical manifestations ordinarily associated with atherosclerosis.

None of the patients except the diabetics were given a special diet. The patients in the treated group were given 500 mg of ascorbic acid orally three times a day but were otherwise the same as the control group.

Table 2. Results of Serial Arteriography in Controls

Age	Diagnosis	Changes in Plaques	Symptom Changes
#1 Age 72	Severe peripheral atherosclerosis	2 plaques bigger; 2 unchanged	Impending gangrene
#2 Age 74	Severe peripheral atherosclerosis	4 plaques bigger; multiple smaller plaques unchanged	No change
#3 Age 68	Diabetes	No change	No change
#4 Age 77	Atherosclerosis, heart disease, diabetes	No change	No change
#5 Age 59	Severe peripheral atherosclerosis	2 plaques bigger	Required amputation
#6 Age 72	Diabetes	No change	No change

Table 3. Results in Group Given Ascorbic Acid (Vitamin C)

Age	Diagnosis	Changes in Plaques	Symptom Changes
#7 Age 69	Severe atherosclerosis, heart disease	3 plaques bigger; 2 unchanged	Died one month later of pneumonia
#8 Age 59	Severe peripheral atherosclerosis, amputation left leg	2 plaques smaller	No change
#9 Age 72	Peripheral atherosclerosis, impending gangrene	3 plaques smaller; 3 plaques unchanged	Claudication decreased
#10 Age 58	Old myocardial infarction	2 plaques bigger; several unchanged	No change
#11 Age 56	Diabetes, angina	7 plaques smaller; 7 unchanged	A growth smaller, less painful
#12 Age 64	Hypercholesterolemia, old myocardial infarction	6 plaques bigger; 2 unchanged	No change
#13 Age 65	Old myocardial infarction, cerebral thrombosis	5 plaques unchanged	No change
#14 Age 51	Diabetes, old myocardial infarction	1 plaque smaller; multiple unchanged	No change
#15 Age 55	Diabetes	1 plaque smaller; 6 unchanged	No change
#16 Age 53	Angina pectoris	5 plaques smaller; multiple unchanged	Much less angina

From these pictures this preliminary study determined that in human test subjects with cardiovascular disease, 60 percent benefited while none of the controls saw an improvement in his condition.

Of the CVD patients given 1,500 mg of vitamin C daily, 70 percent (7 out of 10) improved or their arterial diameters stayed the same. That is, their plaques were reduced (6 out of 10) or unchanged (1 out of 10). Plaques grew in only 30 percent (3 out of 10) of the patients given the 1,500 mg of vitamin C daily. The diameters remained constant or increased in the controls who did not receive the vitamin C.

Willis was well ahead of his time in this arteriography technique. However, other medical scientists showed little interest in the Willis experiments, and these promising results were not followed up on by other researchers.

Notably, these studies were of low dosages, no more than 1,500 mg per day. Thanks to Linus Pauling, we can infer that higher dosages, from 6,000 to 18,000 mg, of vitamin C daily with the addition of lysine are likely to generate even better results.

This experiment remains one of the few studies published in a medical journal on the effect of vitamin C on the lesions of cardiovascular disease in humans. *Inexplicably, since the 1950s no article favorable to vitamin C and its connection to atherosclerosis has appeared in a major medical journal that is widely read by medical doctors.*

Ascorbic Acid Content of Human Arterial Tissue (Willis, G.C., Fishman, S.J., April 1, 1955, Vol 72, pp 500-503)

By 1955 Willis had become convinced that the vitamin C deficiency disease scurvy played a pivotal role in the disease process by disturbing the ground substance surrounding the arteries. As remarkable as the serial arteriographic photographs were, they were apparently not worth a thousand words to the medical community, so Willis thought of another method by which to prove his case. He decided to study the vitamin C content of human tissues.

In their 1955 paper, *Ascorbic Acid Content of Human Arterial Tissue*, Willis and Fishman measured the vitamin C in human aortas (large arteries near the heart) under various circumstances and in this way studied the metabolism of the arterial ground substance in relation to its vitamin C content.

What they learned is summarized in Tables 4 and 5. The values of vitamin C (ascorbic acid) in the arteries in Table 4 are the levels that may be found in sudden death from unnatural and violent causes. In comparison, note that the ascorbic acid content in the arteries of patients who died more slowly after various illnesses as shown in Table 5 is for the most part considerably lower. In 7 of the 20 cases in the hospitalized patient group, no vitamin C was found in the aorta. Willis pointed out that in the older age groups the depletion tended to be particularly marked.

Table 4. Ascorbic Acid Content of Aorta in Cases of Sudden Death

Diagnosis	Hours After Death	Sex	Age	Ascorbic Acid Content of Aorta
#1 Fatal crash injury of the neck	4	M	25	300
#2 Electrocution	10	M	23	200

#3 Drowning	3	F	25	70
#4 Fractured skull and ribs	12	M	45	160
#5 Fractured skull	10	M	45	350
#6 Poisoning	18	M	48	70
#7 Cause unknown	12	M	49	150
#8 Cause unknown	4	M	50	50
#9 Carbon monoxide	4	F	51	140
#10 Sudden death from MI (myocardial infarction)	2	M	56	120
#11 Sudden death from MI (myocardial infarction)	6	M	60	100
#12 Sudden death from cerebral hemorrhage	5	M	65	90

Table 5. Ascorbic Acid Content of Aorta in Cases of Death After Hospitalizations

Case	Hours After Death	Sex	Age	Aorta
#18 Female disease	7	F	18	70
#19 Staph infection	15	M	19	20
#20 Tumor	9	F	19	100
#21 Chronic kidney disease	13	M	23	60
#22 Diabetes/Staph infection	20	F	41	70
#23 Carcinoma	16	M	55	0
#24 Carcinoma	12	M	57	100

Bronchial | 15 | M | 59 | 0 |
#26 Myocardial infarction	12	F	59	0
#27 Carcinoma	15	M	60	0
#28 Hypertension	12	F	62	130
#29 Carcinoma	41	M	63	110
#30 Diabetes, hypertension	26	F	64	60
#31 Thrombosis carotid artery	—	M	66	20
#32 Chronic kidney disease	6	M	66	50
#33 Staph infection	13	F	76	0
#34 Thrombosis carotid artery	7	M	80	0
#35 Cerebral	2	M	80	40
#36 Carcinoma	4	M	82	0
#37 Bronchial	11	F	87	0

Willis and Fishman came to the following conclusions:

· A gross and often complete deficiency of vitamin C (ascorbic acid) frequently exists in the arteries of apparently well-nourished hospital autopsy subjects. Old age seems to accentuate the deficiency. The ascorbic acid depletion is probably not nutritional, but rather is related to the stress of the fatal illness.

· A localized depletion often exists in segments of arteries susceptible to atherosclerosis for reasons of mechanical stress. Adjacent segments, where mechanical stress is less, tend to have a higher ascorbic acid content and atherosclerosis here is rare.

· The significance of this vitamin C (ascorbic acid) depletion lies in the fact that scurvy in guinea pigs results in the rapid onset of atherosclerosis. Furthermore, it has been reported that the aorta can synthesize cholesterol, and the incorporation of radioactive acetate into cholesterol in tissues is reported to be several times more rapid in tissues depleted of ascorbic acid.

· Ascorbic acid deficiency in arteries with resulting ground substance depolymerization may account for the release of glucoprotein that has been noted in the blood of subjects with severe atherosclerosis.

· Preliminary studies suggest that it is possible to replenish the vitamin C in arteries with ascorbic acid therapy.

The Reversibility of Atherosclerosis (G.C. Willis, Canad.M.A.J., July 15, 1957, Vol 77, pp 106-109)

Willis and his team then conducted their landmark experiment in the guinea pig. They knew how to induce atherosclerosis in guinea pigs by depriving the creatures of optimal vitamin C. The scorbutic diet resulted in 100 percent of the subjects succumbing to the disease process. With this knowledge, they wondered whether early plaques could be reversed in the guinea pigs with the induced atherosclerosis. The answer turned out to be, "Yes!"

In the final landmark experiment by Willis, *The Reversibility of Atherosclerosis*, 77 male and female adult guinea pigs were rendered scorbutic in the manner of the first study. The animals were divided into several groups and all groups were fed identical diets except for the vitamin C.

At first, vitamin C was restricted in all groups. A representative group was sacrificed first and every guinea pig in this group was found to have atherosclerosis as expected, confirming the hypothesis that all guinea pigs had the disease.

After intervals of from 21 to 30 days, 50 of these animals were given vitamin C therapy and the remaining 27 were sacrificed. Ascorbic acid therapy consisted of a single intraperitoneal injection of 75 mg of sodium ascorbate, followed by the liberal addition of ascorbic acid powder to the basic scorbutic diet. The animals in this treated group were then sacrificed at intervals of time varying from 1 to 27 days.

Table 6. Number of Animals in Various Experimental Groups With and Without Atherosclerosis

Experiment	Total Animals	With Atherosclerosis	Without Atherosclerosis
Scorbutic diet 42 days with vitamin C (ascorbic acid) from the beginning	12	0	12
Scorbutic diet for periods from 21 to 30 days	27	11	16
Scorbutic diet for 21 to 30 days, then ascorbic acid from 1 to 5 days	25	9	16

Scorbutic diet for 21 to 30 days, then ascorbic acid for 7 to 27 days	25	7	18

When vitamin C was given to scorbutic guinea pigs, the early atherosclerotic lesions resorbed quickly. The advanced lesions were more resistant to reversal, apparently because of the islands of lipids whose only contact with the resorbing process was at the surface. A correlation was made between atherosclerosis in the scorbutic guinea pig and that observed in humans.

Willis wrote in the discussion section that atherosclerosis develops rapidly in guinea pigs without cholesterol feeding, making these animals ideal subjects to study its reversal. In studies with other animal subjects the animals must be given a high-cholesterol diet to induce atherosclerosis. "The hypercholesterolemia that needs to be induced" in subjects that produce their own vitamin C "persists and actually causes atherogenesis to proceed, even when cholesterol feeding is stopped."

Recent Developments

The Willis experiments clearly showed that the atherosclerotic condition will arise in 100 percent of vitamin C-deprived subjects that don't make their own vitamin C, and that atherosclerosis can be reversed. It would be a mistake to perform similar experiments in animals that, unlike the guinea pig, have the capacity to make their own vitamin C. Experiments in mice, rats, or rabbits have limited relevance because vitamin C cannot be eliminated from the bloodstream of the controls. These animals make vitamin C 24/7 in the liver and may not be deprived of it. Furthermore, high cholesterol is required to induce the pseudo-lesions in most other animals.

A new strain of mice that cannot manufacture their own vitamin C has been genetically engineered, and this new mouse strain can develop cardiovascular disease. If the new strain is not given vitamin C supplements, rapid atherosclerosis appears, particularly of the aorta.

A study of 1,605 randomly-selected men in Finland ages 42 to 60 years was conducted between 1984 and 1989 and reported in the *British Medical Journal.* None of the men had evidence of preexisting heart disease. After adjusting for other confounding factors, men who were deficient in vitamin C had three-and-a-half times more heart attacks than men who were not deficient in vitamin C. The scientists' conclusion was, "Vitamin C deficiency, as assessed by low plasma ascorbate concentration, is a risk factor for coronary heart disease" (*British Medical Journal,* Vol 314, Issue 708, 1997).

Another study by Joseph Vita, M.D., an associate professor of medicine at Boston University School of Medicine, found that vitamin C improves blood vessel dilation in patients with coronary artery disease. Vita found that 2,000 mg of vitamin C (which is approximately 30 times the United States RDA) opens arteries by nearly 10 percent — more than some medical treatments.

In another recent study, Thomas Heitzer, M.D., and his colleagues at the University of Freiburg, Germany, compared blood flow in the forearm brachial arteries of 10 healthy male nonsmokers and 10 male chronic smokers after infusing two chemicals, followed by injections of vitamin C. The study published in the journal *Circulation* showed that vitamin C injected into the bloodstream "almost completely reverses endothelial dysfunction in chronic smokers."

It has been widely recognized for at least a decade that endothelial lesions (damage to the walls of blood vessels) are a necessary precondition for the development of atherosclerotic plaques. Oxidized LDL cholesterol and vitamin deficiencies have been theorized to cause these lesions.

According to the American Heart Association, injured endothelial cells may initiate an inflammatory response that leads to increased deposits of "bad" cholesterol and other substances in the artery wall, a process known as atherosclerosis, which can lead to coronary heart disease and heart attack.

In 2006, Board-certified cardiologist Thomas E. Levy, M.D., J.D., analyzed all 27 known risk factors for cardiovascular disease and was able to reduce them all to a single root cause — low vitamin C. Dr. Levy then found more than 650 supporting studies from the scientific and medical literature that are similar to those just mentioned.

Dr. Levy's superb analysis of the risk factors along with the compiled evidence led to the writing of his recent book, *Stop America's #1 Killer*, published by http://www.livonbooks.com. Dr. Levy clearly explains how vitamin C (or the lack thereof) is involved in every facet of heart disease. He blames localized areas of scurvy — focal scurvy in tissues such as the arterial wall where the vitamin C shortage can become acute, even if the deficiency is not as acute in the rest of the body. The Levy book is written by a trained cardiologist *for* cardiologists. It provides an in-depth, technical description of the disease process.

The Future

Cardiologists learn from sources they trust that there is no relationship between vitamin C intake and heart disease and that it is quackery to suggest otherwise. This assertion seems justified because reports of studies are lacking. Yet as vitamin C expert and pharmacological professor Steve Hickey points out, any cardiologist could have performed studies with his own patients from his "petty cash."

Dr. Rath has since published his observation that the Willis findings can be generalized. Most animals make vitamin C and these animals rarely suffer from the same disease process. The few species that require vitamin C to exist will suffer from cardiovascular disease.

The knowledge that heart disease is merely a form of scurvy has been suppressed ever since the series of Willis articles was published in the *Canadian Medical Association Journal* in the early 1950s. Although medicine has for some reason avoided running studies to test the Pauling theory directly, the evidence is mounting in favor of its plausibility and correctness.

Those who argue that "there isn't any evidence" connecting a shortage of vitamin C and heart disease are simply wrong. Those who argue that vitamin C is "harmful to heart patients" have no supporting clinical evidence.

We conclude that maintaining optimal vitamin C levels in the bloodstream is safe and likely prevents cardiovascular disease. One unanswered question then remains: "Why wasn't George C. Willis, M.D. nominated for a Nobel Prize in medicine?"

"There are more than ten thousand published scientific papers that make it quite clear that there is not one body process (such as what goes on inside cells and tissues) and not one disease or syndrome (from the common cold to leprosy) that is not influenced (directly or indirectly) by Vitamin C." — Emeritus Professor Emanuel Cheraskin, M.D., Ph.D., 1974

Chapter 3

Chronic Scurvy

The leading cause of death in the United States is generally called *heart disease*. The disease process is characterized by scab-like buildups that slowly grow inside the walls of blood vessels. Eventually the blood supply to the heart and other organs is diminished from constricted blood vessels, and this results in angina ("heart cramp"), heart attack, and/or stroke.

The better terminology for this disease process is *chronic scurvy*, a subclinical form of the classic vitamin C deficiency disease. The primary symptom of chronic scurvy is atherosclerosis. If the term 'chronic scurvy' were in common use, the entire problem would be solved, as doctors would simply correct the vitamin deficiency.

Scurvy

The devastating symptoms of scurvy, expressed in the wasting and disintegration of the tissues of the body, suggested a large and ubiquitous presence in the body for the factor in nutrition we know today as vitamin C.

Fortunately the disease yielded to the simple therapy of supplying a small ration of the foods that contain the vitamin. The therapy worked its cure long before the vitamin was identified and still longer before its biochemical role began to be as well understood as it is today. — **Linus Pauling, 1986**

Scurvy is the name for the classic vitamin C deficiency disease. *Homo sapiens*, like a few species including the guinea pig, the fruit bat, and a few high order primates, cannot synthesize vitamin C because of a missing enzyme. These species must obtain vitamin C in the diet or die of scurvy. A mere 10 mg of vitamin C will prevent acute scurvy in humans, which results in the long-held hypothesis that ascorbic acid is a vitamin that is only required in minuscule amounts. Persons who don't consume any vitamin C will sicken and die a terrible death.

It is also true that persons who fail to consume any of 39 essential substances, i.e. chemicals the body requires but cannot make, will eventually grow sick and die. These substances are the essential amino acids, known vitamins, minerals, and trace elements.

According to Linus Pauling, "The onset of scurvy is marked by a failure of strength, by depression, restlessness, and rapid exhaustion on making an effort. The skin becomes sallow or dusky. The patient complains of pain in the muscles. He is mentally depressed. Later his face looks haggard. His gums ulcerate, his teeth drop out, and his breath is fetid. Hemorrhages of large size penetrate the muscles and other tissues, giving him the appearance of being extensively bruised. The latter stages of the disease are marked by profound exhaustion, diarrhea, and pulmonary and kidney trouble, leading to death." Without vitamin C in the diet, death occurs in about one month.

Researchers at Arizona State University have discovered that scurvy is prevalent even today and largely undiagnosed. "Overall, we found, 12 percent of Americans had vitamin C deficiency," said J. S. Hampl. "Normally, doctors and other health professionals think of scurvy as a disease of the past, but our research has shown that this really isn't true." Hampl goes on to state:

> Scurvy, or vitamin C deficiency, is associated with low-grade inflammation, fatigue, limping, gum bleeding, or swollen extremities. Vitamin C depletion can also lead to a multitude of other health problems and diseases.
> — **Hampl, et al., 2004**

Vitamin C Was Labeled a Vitamin Before Its Chemical Nature Was Known

Vitamins are substances that we require for life and must obtain in our diet, generally in only small amounts. After the discovery that correcting a vitamin deficiency would cure the corresponding deficiency disease, the hunt to isolate and identify each and every vitamin began.

In the early 1900s the dietary substance that cured scurvy was named vitamin C, even before chemists were able to isolate the substance and determine its molecular structure. Before isolation, scientists assumed that "vitamin C" would have properties similar to the other vitamins that had been isolated and, like the other vitamins, would only be required in small amounts. According to the late science writer Isaac Asimov, writing in 1972:

In 1913, two American biochemists, Elmer Vernon McCollum and Marguerite Davis, discovered another trace factor vital to health in butter and egg yolk. This one was soluble in fatty substances instead of water. McCollum called it "fat-soluble A," to contrast it with "water-soluble B," which was the name he applied to the anti-beri-beri factor. In the absence of chemical information as to the nature of the factors, this seemed fair enough, and it started the custom of naming them by letters. In 1920, the British biochemist Jack Cecil Drummond changed the names to "vitamin A" and "vitamin B," dropping the final e of "vitamine" as a gesture toward taking "amine" out of the name. He also suggested that the antiscurvy factor was still a third such substance, which he named "vitamin C." — **Asimov, 1972**

Isolating and identifying the nature of vitamin C turned out to be a more difficult task for biochemists than isolating the other vitamins that were being identified at the beginning of the twentieth century.

Vitamin C was a different sort of problem. Citrus fruits furnish a comparatively rich source of this material, but one difficulty was finding an experimental animal that did not make its own vitamin C. Most mammals, aside from man and the other primates, have retained the capacity to form this vitamin. Without a cheap and simple experimental animal that would develop scurvy, it was difficult to follow the location of vitamin C among the various fractions into which the fruit juice was broken down chemically.

In 1918 the American biochemists B. Cohen and Lafayette Benedict Mendel solved this problem by discovering that guinea pigs could not form the vitamin.

In fact, guinea pigs developed scurvy much more easily than men did. But another difficulty remained. Vitamin C was found to be very unstable (it is the most unstable of the vitamins), so it was easily lost in chemical procedures to isolate it. A number of research workers ardently pursued the vitamin without success.

As it happened, Vitamin C was finally isolated by someone who was not particularly looking for it. In 1928, the Hungarian-born biochemist Albert Szent-Györgyi, then working in London in Hopkins' laboratory and interested mainly in finding out how tissues made use of oxygen, isolated from cabbages a substance which helped transfer hydrogen atoms from one compound to another. Shortly afterward Charles Glen King and his co-workers at the University of Pittsburgh, who were looking for vitamin C, prepared some of the substance from cabbages and found that it was strongly protective against scurvy. Furthermore, they found it identical with crystals they had obtained from lemon juice. King determined its structure in 1933, and it turned out to be a sugar molecule of six carbons, belonging to the L-series instead of the D-series. It was named "ascorbic acid" (from Greek words meaning "no scurvy"). — **Asimov, 1972, pp 690-700**

After this discovery, ascorbic acid was easily synthesized, and the deadly scurvy became rare.

Flashback: 70 Years Ago

SYNTHETIC VITAMIN PRODUCES STRIKING UNEXPECTED CURES. 'Synthetic vitamin C, called ascorbic acid, in its first actual use on medical patients is producing very striking and unexpected disease conquests,' the British Association for the Advancement

34

of Science was informed by Prof. A. Szent-Györgyi, the Hungarian chemist who played a major role in the artificial manufacture of this important vitamin.

The mouth disorder known as pyorrhea, a certain kind of hemophilia, which is a disease of bleeding, certain forms of hemorrhagic nephritis, and several other diseases against which medicine was helpless are seemingly being cured by ascorbic acid. Ascorbic acid is not a cure for hereditary hemophilia.

'This is the more striking since these pathological conditions have not been thought to be connected with lack of vitamin,' Prof. Szent-Györgyi explained. 'These curative effects suggest that humanity is suffering much more gravely from a lack of vitamin C than has hitherto been supposed.'

Disfiguring colorations of the skin brought on by illness are also made to disappear by ascorbic acid. Patients with Addison's disease, who have a yellow color, can be bleached out again by the use of this substance.
— *Science News,* **Sep 22, 1934**

Is Vitamin C Really A Vitamin?

The common name for vitamin C is ascorbic acid, but this is not the only form of the vitamin. For example, sodium ascorbate is vitamin C in the form of an alkaline salt that is not acidic as are other salts such as calcium ascorbate, magnesium ascorbate, etc. It is the ascorbate fraction of these molecules, called an *ascorbate ion*, which disassociates in solution and contains the vitamin activity.

There is a controversy over whether humans require vitamin C in larger than tiny, vitamin-like amounts. Most plants and animals make vitamin C in ample amounts. On average, mammals synthesize 5,400 mg daily (when adjusted for body weight) and make more of the "vitamin" when under

35

stress. The amount of vitamin C that animals make, adjusted for body weight, is about ten times the amount of coenzyme Q10 (CoQ10) that is synthesized in humans, and roughly 100 times the U. S. Recommended Daily Allowance for vitamin C. Biochemists Linus Pauling and Sherry Lewin cited evidence that about one-half of the vitamin C that we ingest by mouth breaks down and loses its vitamin activity before it enters the cells.

At the cellular level, DNA controls the metabolic machinery in both plants and animals. Primitive cells, bacteria, and even plants are examples of organisms that must be capable of synthesizing more of the chemicals that they require for existence than the organisms capable of movement. The process of evolution has delegated the production of many essential molecules to the DNA of plants. According to Linus Pauling:

> The red bread mold (Neurospora) can live on a very simple medium, consisting of water, inorganic salts, an inorganic source of nitrogen, such as ammonium nitrate, a suitable source of carbon, such as sucrose, and a single vitamin, biotin. All other substances required by the red bread mold are synthesized by it, using its internal biochemical mechanisms. The red bread mold does not need to have any amino acids in its diet, because it is able to synthesize all of them and also to synthesize all of the vitamins except biotin.
>
> The red bread mold owes its survival, over hundreds of millions of years, to its great biochemical capabilities. If, like humans, it were unable to synthesize the various amino acids and vitamins, it would not have survived, because it could not have solved the problem of obtaining an adequate diet. — **Pauling, 1986.**

The forces helping to continue the existence of plants work in opposition to the evolutionary interests of the animals. As Pauling explains, the animals evolved to eat the immovable plants that now make many of the molecules animals require for life.

As the animals began to delegate their cellular manufacturing functions to the DNA of plants, animals became dependent on plants for more and more substances. Plants have had to evolve protections from being eaten into oblivion. Nature is the balance between these two diametrically opposing evolutionary forces.

Generally, the kinds of molecules that animals delegate to plant DNA require many steps to synthesize (introducing greater opportunities for mutation in the animal to interfere with its production), and several vitamins are made of lipid (i.e. fatty) as well as protein components. These complex molecules are generally required by the animals in only small amounts but are found in comparatively large amounts in plants. Many of the essential vitamins are coenzymes, as coenzymes survive the many chemical reactions they facilitate.

Vitamin C as ascorbic acid is unique among the vitamins. First, ascorbic acid is produced by most animals and thus, it is not a vitamin for them. Second, animals produce ascorbic acid in large quantities with respect to their body weight. Finally, ascorbic acid resembles a simple sugar, not a complicated coenzyme. We now know that animals synthesize ascorbic acid using a four-step process. Only a few species have survived to this day having lost the ability to endogenously synthesize ascorbic acid.

This knowledge has spawned an argument about whether vitamin C is a trace factor, the view held by medicine and nutritionists, or whether humans require this molecule in much higher quantities, as Linus Pauling believed.

One of the highest concentrations of ascorbic acid in the body can be found in the adrenal glands, and animals produce more of the "vitamin" when they are under stress. Irwin Stone suggested that a more descriptive term for vitamin C is the "missing stress hormone."

Patrick Holford expands on this version of the "vitamin C isn't really a vitamin" argument and explains how the animals make it:

> Vitamin C isn't a vitamin at all. It isn't a necessary component of diet, at least for all mammals with the exception of guinea pigs, fruit eating bats, the red vented bulbul bird and primates - which includes us. All other species make their own.
>
> This they do by converting glucuronic acid derived from glucose into ascorbic acid ($C_6H_8O_6$). Three enzymes are required to make this conversion. One of these enzymes, or part of the enzyme system, is missing in primates. Irwin Stone proposed, in 1965, that a negative mutation may have occurred in these species so as to lose the ability to produce vitamin C. In primates this is thought to have occurred in the region of 25 million years ago. — **Holford, 1997.**

Holford's argument is not whether vitamin C is required. Patrick Holford, Irwin Stone, Linus Pauling, and others argue that humans suffer a genetic defect and that they require more than tiny, vitamin-like amounts for the best of health.

Ascorbate

Recent authors such as British professors Steve Hickey and Hilary Roberts now refer to all forms of vitamin C simply as *ascorbate*. This term does not connote a tiny, "vitamin-like" trace factor, but just the opposite. A review of vitamin C as well as the basic scientific method can be found in their recent book, *ASCORBATE: The Science of Vitamin C* (2004, http://www.lulu.com/ascorbate). Hickey and Roberts present a new theory in the book, which they call *Dynamic Flow*. This theory has yet to be refuted and fits all known experimental data.

These researchers formulated their theory after a reanalysis of data from the U. S. Department of Health & Human Services' National Institutes of Health (NIH), in which they computed the half-life of vitamin C in the bloodstream to be 30 minutes. In other words, only half of the ascorbate taken by mouth that enters the bloodstream is still in the blood after 30 minutes.

Researchers at the NIH assumed the vitamin C half-life to be greater than one day, and these government researchers recommend only 200 mg of vitamin C daily on that basis.

According to Hickey and Roberts, the flaw in the NIH analysis was only considering vitamin uptake in white blood cells. These vitamin C-hungry cells are the first cells in the body to acquire large quantities of the vitamin from the blood and they quickly become saturated. However, these cells are not representative of the vitamin C requirement and uptake of most cells in the human body.

Hickey and Roberts brought their different calculations and conclusions from the same NIH data to the attention of the

NIH scientists and members of the Food and Nutrition Board, which establishes the RDAs. However, their criticisms fell on deaf ears, and these authorities failed to respond to the scientific criticism. This perversion of the scientific method led Hickey and Roberts to write a follow-up book about the Government's Recommended Daily Allowance of vitamin C. Their book, *The Ridiculous Dietary Allowance* (2005, http://www.lulu.com/ascorbate), expands on these issues.

An important result of the Dynamic Flow theory is the prediction that optimal blood levels can be maintained by taking at least 500 mg of vitamin C every four hours. The 30-minute half-life in the bloodstream is for high dosages and should not be confused with the half-life of the vitamin within cells and tissues, which according to Lewin is 28 days (Lewin 1976).

Vitamin C is now among the most studied substances in the history of science. We know what vitamin C is, we know what it does, and we know how it is naturally synthesized within plants and animals.

We know that ascorbate and ascorbate alone prevents scurvy in the few animals that do not produce it. This has been exhaustively studied in the guinea pig by Ginter (1982) and others.

Only ascorbate has been shown to have a strong effect in preventing and shortening the duration of the common cold. Early experiments found that bioflavonoids provided no additional benefits against colds, either alone or when added to ascorbate (Pauling 1986).

Ascorbate, as the salt sodium ascorbate, can be injected intravenously to control and inactivate viral infections. This curative effect is well documented in Dr. Thomas Levy's book,

Vitamin C, Infectious Diseases, and Toxins: Curing the Incurable (2002).

Intravenous sodium ascorbate recently made news based on reports by Mark Levine, M.D., of the National Institutes of Health and others that it can kill cancer cells at high blood concentrations (Padayatty 2006).

Ascorbate by itself can detoxify the body of heavy metals, including the toxic metal mercury (Levy 2002).

Vitamin C as ascorbic acid and vitamin E cut intensive care unit deaths in half in a randomized, placebo-controlled trial (Nathans 2002).

Recently, medical researchers at John Hopkins University discovered that ascorbate supplements and vitamin E significantly reduced the risk of Alzheimer's by 78 percent and provided an 88-percent reduction in cognitive dementia (Zandi and others, 2004).

Other recent findings, according to the nonprofit Vitamin C Foundation, include the ability of vitamin C as ascorbic acid to reduce the risk of stroke and cataracts and to extend life. Women who took vitamin C supplements for at least ten years proved only 23 percent as likely to develop cataracts as women who received the vitamin only in their diet (Jacques 1997, Mares-Perlman).

In 2000, Yakoyama reported that the risk of stroke was 70 percent higher among those in the lowest quartile for serum vitamin C than among those in the highest. This strong association was confirmed in one of the largest studies to ever examine the issue and published in 2008. University of Cambridge researchers measured vitamin C levels in more than 20,000 people, who were then followed for roughly a decade. When the participants were divided into four groups based on vitamin C levels, those with the highest concentrations of the vitamin in their blood were found to have a 42-percent lower stroke risk than those with the lowest concentrations. The association was seen even after the researchers adjusted for a wide range of stroke risk factors. Lead researcher Phyo K. Myint, MRCP, suggests that the observed increase in risk among people with the lowest vitamin C levels could have clinical implications.

In 2003, Fletcher, et al. reported that low blood vitamin C (ascorbic acid) concentrations in the older British population strongly predict increased mortality. Other vitamins had no effect on mortality. In fully adjusted models there was no evidence for an influence of alpha-tocopherol (vitamin E), beta-carotene, or retinol (vitamin A) on total mortality.

In 1998, Furumoto and other researchers in Japan artificially decreased age-dependent telomere shortening by 52 to 62 percent over untreated control with an enrichment of intracellular vitamin C. Another name for vitamin C could be the anti-death vitamin.

Ascorbic acid has been shown to inhibit the *HMG CoA reductase* enzyme which controls the production of cholesterol. In 1986, the Harwood experiments showed that vitamin C also inhibits the same enzyme, making the vitamin a natural and safe *statin*.

In 2003, a 15-year study of 85,000 nurses at Harvard was published by Osganian. These researchers found that a daily vitamin C pill as ascorbic acid reduced heart disease by nearly 30 percent. Interestingly, dietary intake of vitamin C from food alone seemed to have little effect on coronary heart disease risk, but if women used vitamin C supplements their risk was reduced by 27 percent. According to the numbers in the Harvard study, a 360-mg vitamin C pill daily would save more than 300,000 lives per year.

Arguably, the most important effect of a chronic vitamin C deficiency as is commonly present in most humans is the development of the condition medically referred to as atherosclerosis, which then leads to occlusive cardiovascular disease.

Atherosclerosis is a Symptom of Chronic Scurvy

> Vitamin C is essential for the building of collagen, the most abundant protein built in our bodies and the major component of connective tissue. This connective tissue has structural and supportive functions which are indispensable to heart tissues, to blood vessels, - in fact, to all tissues. Collagen is not only the most abundant protein in our bodies, it also occurs in larger amounts than all other proteins put together. It cannot be built without vitamin C. No heart or blood vessel or other organ could possibly perform its functions without collagen. No heart or blood vessel can be maintained in healthy condition without vitamin C. — **Roger J. Williams, discoverer of the B vitamin, pantothenic acid, 1971**

If we don't eat vitamin C, we die. If a frank vitamin C deficiency disease causes scurvy within 28 days, is there such a thing as a chronic vitamin C deficiency disease that doesn't kill us? Willis thought so. Pauling and Rath and now others think so as well.

The Willis papers referenced Wolbach as the person who had previously demonstrated that scurvy is a disease of the *ground substance*. The ground substance is not made up of cells; it surrounds cells. Willis wrote that it had been known for many years (circa 1953) that stress localizes the lesions of scurvy. "Experimentally, the lesions of scurvy may be focused by the stress of compression or stretching."

The central message of this book is that the leading killer in the United States — the condition that those in medicine call heart disease or occlusive cardiovascular disease — is a low grade scurvy caused by the compression or stretching of arteries near the heart. The disease process progresses fastest in blood vessels under stress that contain the lowest levels of vitamin C. This assertion is based on a surprisingly large body of research that has somehow escaped the attention of the authors of medical textbooks. There is no data from repeatable scientific experiments which refutes this assertion.

The MEDLINE database abstracts available to the general public do not contain older studies or alternative medical studies. Yet, this online database does provide support for the notion that such a thing as chronic scurvy exists.

In one remarkable MEDLINE abstract, University of Chicago authors propose their hypothesis that Lp(a) is a surrogate for vitamin C in "latent scurvy." Because of the publication blackout of unorthodox science as described in The Pauling Therapy Handbook, Volume II, the Chicago authors were unaware of how remarkably similar their thesis was to the previous work of Linus Pauling and Matthias Rath:

Latent scurvy is characterized by a reversible atherosclerosis that closely resembles the clinical form of this disease. Acute scurvy is characterized by microvascular complications such as widespread capillary hemorrhaging. Vitamin C (ascorbate) is required for the synthesis of collagen, the protein most critical in the maintenance of vascular integrity. We suggest that in latent scurvy, large blood vessels use modified LDL - in particular lipoprotein(a) - in addition to collagen to maintain macrovascular integrity. By this mechanism, collagen is spared for the maintenance of capillaries, the sites of gas and nutrient exchange. The foam-cell phenotype of atherosclerosis is identified as a mesenchymal genetic program, regulated by the availability of ascorbate. When vitamin C is limited, foam cells develop and induce oxidative modification of LDL, thereby stabilizing large blood vessels via the deposition of LDL. The structural similarity between vitamin C and glucose suggests that hyperglycemia will inhibit cellular uptake of ascorbate, inducing local vitamin C deficiency.

Readers of this book will learn that the miniscule amount of vitamin C in the U. S. Government's RDA of 60 to 75 mg will prevent death from the vitamin C deficiency disease scurvy. However, the vitamin C RDA is not sufficient for the body to maintain the strength and integrity of its coronary arteries that lie close to the heart. The amount of daily vitamin C necessary to prevent and treat latent or chronic scurvy, also known as human atherosclerosis, is more than 100 times higher than the RDA. Furthermore, biochemists Linus Pauling and Sherry Lewin cite evidence that only half of the vitamin C taken orally reaches cells. The other half breaks down and loses its vitamin activity during digestion or in the bloodstream before entering the cells.

Among its many metabolic functions, vitamin C as ascorbate is required in the manufacture of the protein collagen (Pauling 1986). Collagen provides strength and structural integrity to animal tissues. The role of collagen is described more fully in the next two chapters.

The ongoing repair and maintenance of tissue induces a daily need for new collagen. In human scurvy, the body disintegrates from a lack of collagen in approximately one month. It takes longer for those getting some vitamin C, but less than 10 mg, to succumb to a slower form of scurvy, generally about 150 days.

It is wise to heed Linus Pauling's advice that one should not stop taking vitamin C, "not even for a single day."

Chronic Scurvy Verified by CardioRetinometry ®

There has been an exciting new development in Britain. Optometrist Sydney Bush has noticed that the microscopic retinal arteries in the eye convey information about a person's cardiovascular health.

Atheroma, or a soft, white plaque formation, is the name that eye doctors give to abnormal formations that appear in arteries. Atheromas in the microscopic arteries of the retina have been clearly visible to eye doctors, though until recently they did not believe that such buildups were reversible. Optometrists have noticed that high blood pressure soon follows the formation of these anomalies.

Dr. Bush accidentally discovered that early atheromas, those that have not yet calcified, can be reversed in those patients instructed to take from 3,000 mg to 10,000 mg of vitamin C. Dr. Bush made his discovery while studying eye infections in contact lens wearers. Vitamin C was being tested as a preventive measure for these infections and, serendipitously, Dr. Bush noticed that atheromas began to disappear, but only in the study patients who were taking vitamin C.

Dr. Bush agrees with the long-held view that human arteries weaken without vitamin C and other necessary nutritional support. He reports that some patients require as much as 10,000 mg daily to reverse soft atheromas.

Bush has invented a new diagnostic technique called *CardioRetinometry®* that can be used to monitor atheromas. He believes this new non-invasive method of analysis will help to revolutionize the diagnosis of occlusive cardiovascular disease.

Dr. Bush reports that he has increasingly noticed the vitamin C effect since 1999 using CardioRetinometry® in the Hull Contact Lens and Eye Clinic. "Such a discovery requires urgent evaluation," he says.

Dr. Bush not only agrees that chronic scurvy exists, but he says that it can be accurately measured. Eye doctors can now be trained to diagnose this condition by examining the microscopic arteries behind the eye before any symptoms of heart disease manifest.

Dr. Bush's discovery holds promise that vitamin C will reverse the condition in short order at the optimal dosage determined by CardioRetinometry®. Dr. Bush states, "People today are under the seriously mistaken impression that nobody dies of scurvy anymore! These studies may prove that we are all dying faster from scurvy than hitherto suspected."

The pericorneal vasculature that is frequently studied by contact lens practitioners shows that scurvy affects all humans some of the time and most humans most of the time.

"The largely unrecognized chronic subclinical form of scurvy can best be diagnosed (and easily cured) by optometrists using sequential electronic retinal artery images and variable amounts of vitamin C, occasionally with other nutrients," according to Dr. Bush.

CardioRetinometry® clearly demonstrates the relationship between vitamin C intake and "atheromas," plaques which form on the arteries that serve the retina in the eye. Dr. Bush has published "before" and "after" pictures taken with his new method and he strongly advocates the need for rigorous studies.

> The atheroma of the retinal arteries is a virtually perfect surrogate outcome predictor of coronary heart disease and will continue to be so as long as the eyes are connected to the rest of the system. The modern electronic eye camera/microscopes with high definition magnification facility show the impacting of the cholesterol beautifully and also its redissolving into the bloodstream when the system is restored to balance. And this is seen in arterioles too small to be seen with the naked eye! — **Sydney Bush**

While day-to-day variations in the pericorneal vessels are a 'barometer' of 'ephemeral' scurvy, especially when viewed via the slit lamp biomicroscope of the contact lens practitioner, little attention has been paid to it. Bush goes on to state:

> The pericorneal arterioles and capillaries can and are graded in my system of practice into ten degrees of scurvy allowing the accurate prediction to patients of how much or little vitamin C they have been ingesting. The highest mark anybody has had is 94%. When I started this grading, c.1997, I confounded my nursing staff by being able to correctly identify patients who ate no or few greens.
>
> But the same ease of observation does not attach to identification of the chronic subclinical variety. It cannot identify dietary faults in the most recent past. Like the slow buildup of vitamin E in the body fat and cell walls of the brain, it takes over a month to be sure what is happening to the cholesterol in the retinal arteries.

Dr. Bush has even found evidence that calcified, "hard" plaques can be reversed over the course of two years on a high vitamin C intake.

The moral of the story is that one should undergo regular examinations of the retinal arteries conducted by a suitably equipped optometrist trained in CardioRetinometry®. In Dr. Bush's opinion, this noninvasive safeguard can help protect one's cardiovascular health and probably many other systems of the body, as systems do not act in isolation.

Natural Vitamin C and the Vitamin C Complex

Recently, advocates of a so-called "natural Vitamin C complex" have challenged the idea that ascorbic acid is really vitamin C. This has resulted in a number of well-intentioned but misguided people among the alternative medical community. There is no science to support the notion that a natural vitamin C complex even exists, much less that it is the "real" vitamin C.

The dietary substance that causes scurvy when missing and cures scurvy when present is, by definition, vitamin C. Linus Pauling was unequivocal in his belief that the ascorbate fraction of ascorbic acid (called the ascorbate ion) is vitamin C. Referring to scurvy in his landmark book, *Vitamin C and the Common Cold* (1970), Pauling stated, "Ascorbic acid is an essential food for human beings. People who receive no ascorbic acid (vitamin C) become sick and die."

There is a growing school of thought that vitamin C isn't vitamin C among the unlikely foes of Linus Pauling — the natural purists who proclaim that only vitamins gleaned from plants are true vitamins. The view of these alternative healers as summarized by authors Thomas S. Cowan, M. D., and Sally

49

Fallon in their recent book, *The Fourfold Path to Healing* (2004), is that the real vitamin C is "actually a complex of nutrients that includes bioflavonoids, rutin, tyrosine, copper and other substances known and unknown" (Cowan and others, 2004, p 21). Ascorbic acid, which has been considered by scientists to be vitamin C since at least 1937, has only a "supporting" role, according to Cowan and Fallon, who write that ascorbic acid is only present in plants "as a preservative for this complex, serving to keep it together in the plant tissue, preserving its integrity, freshness and color" (Cowan and others, 2004, p 21). Cowan and Fallon even go so far as to say that "ascorbic acid is not a food for us; that which it preserves is our food" (Cowan and others, 2004, p 21). The naturalists assert that too much "synthetic" ascorbic acid is harmful, especially when not accompanied by the vitamin C complex. If the naturalists are right about the C complex being the true vitamin C, then Willis, Pauling, Lewin, Stone, Cheraskin, Levy, Hickey, Roberts and all other scientists are wrong, as are the now more than 100,000 reviews and analyses from more than 70 years of vitamin C science.

There is massive scientific support for Linus Pauling's position that ascorbic acid (the ascorbate ion) is vitamin C. No scientific basis could be found for the notion that a C complex exists or can cure scurvy without ascorbic acid present. This assertion is proven every day in hospitals around the world. Comatose patients are kept alive using only ascorbic acid. There are no hospitals keeping patients alive on a feeding tube with a vitamin C complex.

Plant complexes containing bioflavonoids may have health benefits, however. Linus Pauling himself advised eating a wide variety of foods because there is a chance that not all the molecules that a healthy body requires have been discovered. However, plants have generally evolved to avoid being eaten.

The most frequently mentioned health benefit of the bioflavonoids, such as the quercetin and rutin that are commonly found with vitamin C in the white of the rinds of oranges and mentioned in the literature, is that these nutrients strengthen the walls of tiny capillaries. However, there is no evidence that any molecule other than ascorbate, the same molecule produced by the liver of most animals, can replace vitamin C to prevent scurvy or provide equivalent metabolic properties.

Fortunately for humanity, synthetic bioidentical vitamin C is inexpensive and offers the hope of better health to everyone. Vitamin C Foundation researcher Ralph Lotz points out that the 100 mg of the "natural vitamin C complex" sold by one company is 1,315 times more costly than ordinary vitamin C.

Chapter 4

Lp(a)

If you are human and reading these words, then you are alive but you probably cannot make vitamin C in your body. Therefore, you owe your continued existence to a form of cholesterol in the blood that was discovered in 1962 by Blumberg, et al. This peculiar form of *low density lipoprotein (LDL)* is called *lipoprotein(a)*, "small a," or simply *Lp(a)*. Lp(a) is not to be confused with *lipoprotein(A)*, "large A," another lipoprotein which often appears on lab reports. They bear no relation to each other.

Lp(a) is an ordinary cholesterol particle with a sticky *apoprotein* particle called *apo(a)* attached to its surface. Lp(a) is formed in the liver. Lp(a) does not come in a standardized size or mass; the apo(a) may attach to a variety of low density lipoproteins during its formation. All human beings have the capacity to make Lp(a) and it is found in almost all human blood. However, there can be a thousand-fold range in its plasma concentrations. High levels of Lp(a) are associated with high incidence of cardiovascular disease.

Forty years after the Willis experiments it was discovered that only this one form of cholesterol, Lp(a), begins the process of forming atherosclerotic plaques. By 1990, after this discovery was experimentally verified in Germany, Linus

Pauling and Matthias Rath proposed their new theory which singled out Lp(a) as the most significant and potentially the most dangerous variant of LDL ("bad") cholesterol. According to Dr. Rath:

> High cholesterol, specifically LDL (bad) cholesterol, is not the cause of cardiovascular disease. The newest research tells us that Lp(a) will cause cardiovascular disease ten times more likely than high LDL (bad) cholesterol.
>
> This fact was revealed during a recent reevaluation of the Framingham Heart Study, the largest cardiovascular risk factor study ever conducted. Today Lp(a) has been confirmed as the leading risk factor for many forms of cardiovascular disease.

Lp(a) as a Surrogate for Vitamin C

The Pauling/Rath theory is based on the knowledge that excess Lp(a) causes the plaque buildup on the walls of arteries that leads to chest pain, heart attacks, and strokes. In their view, the peculiar Lp(a) molecule explains a great deal: Lp(a), the friend who may become a foe, substitutes for vitamin C in the few species that do not make the vitamin.

Lp(a) has important health benefits in the absence of vitamin C; it provides many of the same functions in the body that the missing vitamin C would have provided. For example, Lp(a) provides an alternate way to strengthen and stabilize vitamin C-starved arteries. Lp(a) may even be the reason coronary bypass surgeries succeed using tissues from veins that are weaker than the arterial tissue being replaced.

Lp(a), Not Ordinary LDL, Has Lysine Binding Sites

Medical scientists have known for 20 years that Lp(a) in the blood initiates plaque formations called atherosclerosis. Ordinary LDL cholesterol does not. In 1987, the Brown-Goldstein Nobel Prize in medicine of that year first brought the so-called cholesterol binding sites to the attention of an audience larger than a group of medical researchers.

Researchers had discovered that Lp(a) binds to lysine strands that become exposed when the wall of the artery suffers a sore or lesion. Yet for reasons unknown, medical journal editors, researchers, doctors, and the media never picked up on this important finding that Lp(a), i.e. *LDL + apo(a)*, and *only* Lp(a) is the sticky form of cholesterol with the famous lysine binding sites.

The Collagen Triple Helix

Inside the wall of every blood vessel lies the collagen girder shaped into a triple helix. These girders, like steel buried in a concrete highway, provide the arteries with their tensile strength and stability. The amino acids lysine and proline are the major building blocks that our cells use to construct the collagen girders.

Vitamin C is required as a linking molecule during the construction of collagen, and the vitamin is destroyed in the process. The collagen triple helix is living tissue and needs to be replenished periodically. The more vitamin C present, the more collagen the body can produce. If the body can make more collagen and more of its protein cousin elastin, arterial tissues become stronger and more resilient to tiny fractures. Without any of the vitamin, the girders providing support to various tissues begin to break down, and death from scurvy inevitably occurs.

Coronary arteries are squeezed by the heartbeat, which increases the daily requirement for new collagen and thus, vitamin C. When vitamin C is lacking in the cells that manufacture collagen, the arteries tend to deteriorate as the arterial wall weakens. Surface disruptions emerge on the blood vessel wall, especially from wear and tear near the beating heart. Dr. Rath likens the stress to that of stepping on a garden hose thousands of times every day. Cracks develop, especially in the coronary arteries that are subjected to the most mechanical stress from the squeezing and bending as the heart beats.

Blood vessels can be kept strong by an optimal intake of vitamin C (and other antioxidants and amino acids), making it less likely that lesions will develop. According to the theory, if lesions do not develop, artery blockages do not develop.

Arterial lesions are like potholes along our most crowded vascular highways. When a lesion begins to form, strands of lysine and proline from the deteriorating collagen become exposed by the injury. The floating Lp(a) in the bloodstream comes to the rescue; it is attracted to the exposed strands of lysine and proline. As these strands appear in the pothole, the sticky Lp(a) binds with it and forms a patch, unless something happens to make the Lp(a) unattractive. The Pauling/Rath invention, *Lp(a) binding inhibitors*, is the subject of Chapter 6.

The knowledge of how plaques form from Lp(a) in the blood led Linus Pauling and Matthias Rath down the road that resulted in their *Unified Theory*. Lp(a) was unknown at the time of Willis. The Pauling and Rath experiments on guinea pigs were similar to the Willis experiments, though this time Lp(a) was measured to test their theory. Not only did these confirming experiments find Lp(a) (its precursor apo(a)) in the guinea pig, but serum Lp(a) levels only rose in the animals on a reduced vitamin C diet.

The News Report, "Lipoprotein Ups Heart Attack Risk"

On September 4, 2000 there was a brief news report that would have sent shock waves through the entire medical-pharmaceutical industry had it not been downplayed. Researchers at Oxford University were about to publish their paper in the American Heart Association Journal *Circulation* with their findings that people with high levels of lipoprotein(a) are 70 percent more likely to have a heart attack than those with lower concentrations.

Lipoprotein Ups Heart Attack Risk

09/04/2000 Press Release

DALLAS — Cardiac patients with high levels of a little-known form of "bad" cholesterol in their blood are 70 percent more likely to have a heart attack than those with lower concentrations, according to a study released Monday.

The obscure cholesterol particle - called lipoprotein(a) - is especially insidious because it's difficult for doctors to measure reliably and because its levels have little to do with the better-known form of "bad" cholesterol, called LDL.

'This study suggests there is a clear association between Lp(a) and an increased risk of heart disease.'

Researchers gathered data used in the study from 27 different studies tracking more than 5,200 people who had heart disease or survived a heart attack. The average age of the people involved in the study was 50.

The number of heart attacks suffered by individuals with the highest Lp(a) concentrations was compared with the number of heart attacks among those with the lowest Lp(a) readings. During a decade of follow-up, the highest group had 70 percent more heart attacks than the low-level Lp(a) study subjects.

Lp(a) was first pinpointed in the blood some 40 years ago, but doctors don't normally screen for this lipoprotein because no standardized screening exists and because even when the Lp(a) is known, very little can now be done to modify it. Unlike other kinds of cholesterol, Lp(a) in the blood is 95 percent determined by genes, so drugs and changes in diet have little effect on it.

While this news was exciting to us at the time, the full news story did contain inaccuracies, or at least misleading statements. Whether those distortions were intentional or not, what follows is an attempt to set the record straight.

For example, there were then more than 1,200 published scientific papers on what the news report called an "obscure" cholesterol particle.

The news report implied that in the previous decade, researchers had not been able to link high Lp(a) to an increased risk of heart attack in the general population. This is belied by the science available from MEDLINE.

The news report claimed that the reason doctors do not normally screen for Lp(a) is that there are no standardized blood tests. This is not true now and it was not true then.

The wire service report repeated the often-made claim that Lp(a) is not alterable with drugs or changes in diet. *Actually, cholesterol-lowering statin drugs are known to elevate Lp(a) at the same time they reduce ordinary cholesterol. In Canada, health authorities require this warning to be in all advertisements for these drugs. The*

FDA does not require this warning, and American doctors are ignorant of this adverse effect.

Finally, the wire services reported that Lp(a)'s exact role in the blood is unknown, completely deflecting the discovery that only Lp(a) has the sticky lysine binding sites.

Routine cholesterol screenings still do not include separate Lp(a) scores. Historically, the Lp(a) numbers have been lumped with the LDL score. At least one statement made in the report is true: **There are no prescription drugs that lower Lp(a) levels**.

Beisiegel

Linus Pauling became interested in Lp(a) after a research group in Germany confirmed that plaque deposits on human aortas were entirely Lp(a) and not ordinary LDL. The young Matthias Rath was a member of this German research team. He traveled to the United States to join Linus Pauling after he realized that Lp(a)-based plaques were an evolutionary countermeasure for chronically low levels of vitamin C in humans.

Another important finding that impressed both Pauling and Rath is that the "sticky" Lp(a) particles have only been found in species that do <u>not</u> make their own vitamin C. Most species make large quantities of vitamin C in the liver or kidney; Lp(a) is not found in their blood, and rarely do these species suffer from cardiovascular disease.

Humans are a rarity among life forms on Earth, in that they must obtain all of their vitamin C entirely from the diet.

Group	Endogenous Vitamin C	Lp(a) in the Blood	CVD
Humans	No	Yes	Yes
High order primates	No	Yes	Yes
Guinea pigs	No	Yes	Yes
Other animals 99.9+% of species	Yes	No	No

Pauling and Rath were also impressed by the experiments and observations of the Canadian doctor Willis, who had first observed that plaques generally form in locations where blood pressures are the highest, i.e. in the coronary arteries near the beating heart. "If cholesterol were the cause and not the effect of heart disease," they wondered, "why aren't plaque formations and the subsequent infarctions more randomly distributed throughout the body?"

The Pauling/Rath Experiment

As readers know by now, most animals produce their own vitamin C and, according to Pauling, do not have Lp(a) in their blood. However, guinea pigs cannot make their own vitamin C either. When these animals are deprived of vitamin C they die a horrible death in a matter of weeks. Pauling and Rath wondered whether they also produce Lp(a) like humans on a low vitamin C diet. An experiment to test the theory was conducted at the Linus Pauling Institute of Science and Medicine.

In the laboratory the guinea pigs were divided into two groups. The animals in one group were given small amounts of vitamin C, a scorbutic diet (roughly equivalent to the U. S. RDA), and serum Lp(a) (apo(a)) was detected. The Lp(a)/apo(a) levels rose over the course of their lifetimes and, as Willis discovered, the animals developed atherosclerosis with lesions similar to those found in humans. The animals in the other group (on the same diet) were given the human equivalent adjusted for body weight of 3,000 to 5,000 mg per day of vitamin C. Those animals did <u>not</u> develop the disease and their Lp(a) levels did <u>not</u> become elevated. The only difference between the two groups was the vitamin C in the pigs' diet. Too little vitamin C made the difference.

The Pauling/Rath experiments correlated Lp(a) with the atherosclerosis caused by a low vitamin C scorbutic diet and were published in the *Proceedings of the National Academy of Science* (Pauling/Rath 1988). This experimental work not only confirmed the earlier findings of Willis but also resulted in the first United States patent for reversing heart disease without surgery in 1994.

The Pauling/Rath *Unified Theory* agrees with Willis and holds that heart disease is the body's healing response to a low grade form of scurvy, a slower form of the disease than that which killed the sailors without fruits on the high seas.

Pauling and Rath, like Willis, explained that scurvy and cardiovascular diseases ultimately develop because tissues lack sufficient collagen to keep them strong. Blood vessels in species that do not make endogenous vitamin C tend to weaken because collagen production is dependent on vitamin C levels. When vitamin C and collagen decline, the body compensates by raising Lp(a). The elevated Lp(a) initiates plaque formations over the blood vessel lesions, making the plaques, in Dr. Rath's words, "nature's plaster casts."

Measuring Lp(a)

The truth is that Lp(a) is still not generally screened for during ordinary cholesterol testing, and special procedures are required for accurate readings. Contrary to the wire news report, there are companies that accurately measure Lp(a), and at a reasonable cost.

One such company, Atherotech of Birmingham, Alabama, offers its VAP® Test kit across the country. The kit includes complete lab instructions that explain how to process the sample before returning it to Atherotech for analysis. Atherotech not only measures Lp(a), but this laboratory breaks Lp(a) down into five subgroups.

Moles Versus Mass

There are two distinct measures of Lp(a) that are used to report the risk, which causes some confusion. The original measure is of *mass per volume* (mg/dl), and the newer measure is of the *number of particles per volume* (nmol/l). The problem the newer measure solves is that Lp(a) molecules come in different sizes. Both the mass and the size of Lp(a) particles vary, and according to Lp(a) experts, smaller particles (less mass) are more atherogenic than larger particles. In other words, small Lp(a) particles tend to form plaques faster. A smaller total mass might be more dangerous; it depends on the size of the particles, and vice versa, higher mg/dl scores may or may not be indicative of the danger of atherosclerosis.

Recently, the solution to this problem has been to report Lp(a) in terms of moles - nmol/l - or the number of particles.

The more particles per the same volume implies that the Lp(a) are smaller and thus more dangerous.

The normal reading for mg/dl is around 20; the normal reading for nanomoles/liter is around 70. Unfortunately, again, according to the Lp(a) experts, there is no way to accurately convert from one measure to another. It is like apples (mass) and oranges (molecules).

The Atherotech VAP® Test was an early attempt at overcoming the problem of reporting Lp(a) in terms of mg/dl. Atherotech's answer to the confusion was to create a standardized score which would provide a better comparison between scores and be a better indicator of risk. The VAP® score is not the mass of Lp(a) per se, but rather the score is the mass of an equivalent number of ordinary LDL cholesterol molecules. The VAP® Test also provides physicians with the mass of the Lp(a) and breaks Lp(a) into five ranges based on mass.

Recent confusion has been caused by the U. S. Food and Drug Administration (FDA), which now allows testing laboratories to estimate Lp(a). The FDA has approved these computations (guesses) because Lp(a) is difficult and costly to measure. According to the announcement, the FDA has decided that because there is "nothing" currently recognized that can alter Lp(a) levels, the cost/benefit does not justify the expense of separating and accurately measuring Lp(a). In our experience, the calculated scores are useless.

Today, because of these new FDA rules, people often see their Lp(a) numbers without knowing how accurate these measurements are. Neither they nor their doctors know if Lp(a) numbers are measured (real) or computed (estimates). One clue is that Lp(a) is a low density lipoprotein and therefore a subset of LDL. If the reported Lp(a) is higher than the total LDL score and in the same units (e.g. mg/dl), then the score is probably an estimate and therefore invalid.

One Nationwide lab that consistently gives very large numbers has refused to answer questions from customers about how its Lp(a) numbers are generated, i.e. whether its numbers are estimated or measured. Given the volume of testing, we can guess which method they are using.

If you are concerned about your heart attack risk, make sure your Lp(a) score is *measured* and not *computed*.

In our opinion, the Atherotech VAP® Test provides "gold standard" Lp(a) measurements. As verification, I personally had my own blood drawn and had the same blood sent to different laboratories. The VAP® number from Atherotech was 3 mg/dl. The hospital lab reported the Lp(a) value as "less than 5 mg/dl." Other fine laboratories that measure Lp(a) are Berkeley Lab and Great Smokies Diagnostic Laboratory (now Genova Diagnostics).

Dr. Pauling, using the mg/dl measurement common in 1994, stated, "If you have more than 20mg/dl of Lp(a) in your blood it begins to deposit plaques, causing atherosclerosis."

Because no prescription drug or low-fat diet has been shown to alter Lp(a), this molecule is generally held to be a genetic factor, which of course it is. However, this is misleading because as shown by Pauling and Rath, Lp(a) can be controlled by dietary supplements in guinea pigs.

According to Dr. Rath, studies show that a special diet (i.e. one that is low in fat) does not generally influence Lp(a) blood levels. However, vitamin C and vitamin B3 (niacin) can lower blood levels of Lp(a) by about 15 to 30 percent. Vitamin B3 probably acts on Lp(a) production in the liver, while vitamin C levels, according to the Pauling/Rath theory, attack the root cause.

Furthermore, significant documentation attests to the fact that vitamins belong among the most powerful agents in the fight against heart disease. Dr. Rath's Foundation states that this finding has been established by studies of thousands of people over many years. Following are the results of major clinical studies:

- Vitamin C cuts heart disease rate almost in half (documented in 11,000 Americans over ten years).
- Vitamin E cuts heart disease rate by more than one-third (documented in 36,000 Americans over six years).
- Beta-carotene (provitamin A) cuts heart disease rate almost in half (documented in 36,000 Americans).

Orthodox Medicine and Lp(a)

Orthodox medicine has researched Lp(a) extensively since Pauling first took his *Unified Theory* lecture tour to various universities in 1989. The National Institutes of Health's MEDLINE database makes it easy to monitor the ever-growing flurry of published research. Before 1992 there was only a handful of MEDLINE study abstracts which mentioned lipoprotein(a) or Lp(a) in the subject or title. However, by 1997 there were more than 1,500 studies, papers, and letters that matched. At last count (in 2008) there were 11,000 Lp(a) papers catalogued (out of 125,000 matching lipoprotein). Although the pioneering work of Pauling and Rath is not cited (and apparently unknown to most authors), these studies generally confirm the importance of Lp(a) as a risk factor in premature coronary vascular disease.

The volume of this research is now massive (the abstracts alone would require approximately 11,000 pages). Following are but two examples of the conclusions from papers published in mainstream medical journals:

Lipoprotein Lp(a) excess has been identified as a powerful predictor of premature atherosclerotic vascular disease in several large, prospective studies. We measured serum Lp(a) and other lipid parameters... Although triglyceride and LDL, HDL and total cholesterol levels were similar in the restenosis and no-restenosis group before PTCA, Lp(a) was significantly higher in the restenosis group. This study provides the first evidence in man of a significant role for lipoprotein Lp(a) in unstable angina.

Lipoprotein(a): a potential biological marker for unruptured intracranial aneurysms. The prevalence of elevated Lp(a) levels in and the high degree of association of raised Lp(a) with the presence of IAs. While the two drugs (pravastatin and simvastatin) caused the expected reduction of plasma total and LDL cholesterol levels, no significant changes in Lp(a) were noted.

These abstracts, along with more than 11,000 other MEDLINE abstracts, can be obtained from the NIH by simply searching for "Lp(a)" in the subject or title beginning with the year 1992.

Although among these thousands of studies a few abstracts make the claim that they could not correlate Lp(a) to coronary arteriosclerosis, the vast majority have found Lp(a) to be a highly atherogenic risk factor, a feat that science was never able to accomplish with LDL alone.

Dr. Matthias Rath was able to publish one study in the *Journal of Applied Nutrition*. He was apparently forced to study his own product formula because it is difficult to interest mainstream researchers in the benefits of vitamin C. (Conducting vitamin C research interferes with a researcher's ability to secure future grant money, as readers will learn in The Pauling Therapy Handbook, Volume II.)

Rath and his team were able to demonstrate that atherosclerotic deposits can be prevented and even eliminated over the course of one year. This improvement was statistically significant compared to the controls whose atherosclerosis increased dramatically. However, it must be said that according to the abstract, the amounts of lysine that were investigated are about 10–20 percent of what one normally obtains in his diet — hardly therapeutic.

It is important to differentiate between these rather low dosages and the larger dosages recommended by Linus Pauling. The Pauling recommendation of 5 to 6 grams of lysine daily is about 30 to 40 times the dosage that was reportedly used in the Rath study.

Small Pilot Study

Even with the growing body of research, no one was studying Lp(a) using the dosages recommended by Linus Pauling. In 1998, after the National Institutes of Health denied the second grant request made by the nonprofit Vitamin C Foundation to study the Pauling therapy, the Foundation conducted a small pilot study to test individual reports that nutritional products can lower Lp(a) levels by more than 30 percent in humans.

The study involved the very first Pauling therapy product sold in the world, manufactured by Tower Laboratories Corporation of Las Vegas, Nevada. The study sought to verify reports that this product, high in vitamin C, lysine and proline, with vitamin E, vitamin A, carnitine, and a few B vitamins, can dramatically lower Lp(a) levels within 18 months. In an effort to reduce costs, those who participated measured their progress and submitted their lab reports to the Foundation.

The decline in Lp(a) from eight subjects who had been given free product was 68 percent. In at least two cases, recorded Lp(a) levels in the blood initially rose approximately 10 to 20 percent before they began to decline significantly. Dr. Rath has written that he has observed the same effect in total cholesterol. At first, cholesterol actually *increases* from the therapy, soon followed by a steep drop.

However, the study was flawed because the "before" and "after" lab studies were sometimes conducted by different labs and may have had mixed measured results with estimated results (the Vitamin C Foundation was unaware of this particular issue at that time).

Although the pilot study was flawed and never published, there were a few spectacular individual reports:

My cardiologist was floored when I participated in your study some years ago. I participated in a survey that you ran using two jars of [Pauling Therapy formula] per month. My Lp(a) was running around 120. After six weeks, the score was 35. My medical records have a big "Wow!" in the margins. It's still 5 over the norm, but sure is a big improvement. My cardiologist didn't think anything would knock my cholesterol and Lp(a) down until I used *Heart Technology*, since my father died at 47 from a second massive heart attack. Poor genetics can be overcome. Now that I'm in my 50's (Hooray!), I would like to see my 60's and 70's. — **Laura M.**

The Vitamin C Foundation is currently sponsoring another study and its founders hope that there will be more controlled clinical studies conducted by other researchers and laboratories. It would be important to know whether the high doses that Linus Pauling recommended consistently lower the Lp(a) risk in humans, especially since there are no drugs available to lower Lp(a).

There is reason to believe that the addition of proline and/or vitamin A may increase the benefits of vitamin C and lysine. This supposition is based on an as yet unpublished three-year study in the United Kingdom. The Kenton study was conducted in men and reportedly used vitamin C, vitamin E, and lysine at the correct dosages. No proline was used. Atherosclerosis was dramatically reduced compared with controls, but according to the initial report the researchers found no change in Lp(a) among the two groups. (Perhaps the Lp(a) in this study was measured and not calculated.) If confirmed, this study indicates that vitamin C and lysine together, but without proline, do not affect Lp(a).

Another bit of evidence comes from a professor at a New York medical school. He had been taking vitamin C and lysine for years but his Lp(a) was still elevated. He made one change to his supplementation; he added proline. After six months his Lp(a) had dropped by 30 percent, and after 14 months his Lp(a) was zero. After learning the results of the second test, he called me with his report.

"Vitamin C has been under investigation, reported in thousands of scientific papers, ever since it was discovered (circa) fifty years ago. Even though some physicians had observed forty or fifty years ago that amounts a hundred to a thousand times larger (than the RDA) have value in controlling various diseases, the medical profession and most scientists ignored this evidence." — *How to Live Longer and Feel Better,* Linus Pauling, 1986

Chapter 5

The *Unified Theory*

Various theories attempt to explain what causes the cardiovascular disease that leads to heart attack and stroke. There is a cholesterol theory; a fat (saturated and polyunsaturated) theory; the long-neglected homocysteine theory first proposed by Kilmer McCully; an oxidized cholesterol theory; a free radical/heavy metal theory; and even a microbe theory. Every theory attempts to explain what causes the lesion, i.e. the initial injury or crack in the artery that precedes the development of atherosclerosis.

At least one major theory is never mentioned in the medical journals or lay media: *The vitamin C theory.* Linus Pauling and Matthias Rath argue that the great problem of cardiovascular disease is the body's reaction to a chronic rather than acute vitamin C deficiency. The Pauling and Rath *Unified Theory* explains how humans and the few other species that have lost the ability to manufacture the vitamin in their bodies have been able to survive this negative mutation.

A recently-discovered form of cholesterol has properties which allow it to substitute for vitamin C. However, this adaptation is not perfect and still leads to clogged arteries, heart attacks, and strokes later in life.

This common sense idea has not been seriously investigated by modern medicine for 50 years. Yet cardiologists routinely tell their patients that there is no "proven" value in taking vitamin C for heart disease. Technically, this statement may be accurate, but it is misleading. The implication is that experiments have been run which have proven that vitamin C has no value. However, no such experiments have ever been run, or if they have, the results have never been published in a reputable journal.

On the contrary, all research and experiments we know of provide evidence that vitamin C has value, with perhaps one possible exception. Rimm, et al. of Harvard conducted epidemiological, not experimental, studies (meaning the results were from surveys and did not measure blood levels). Rimm concluded "from almost 40,000 subjects" that vitamin C was of no value to subjects who had contracted cardiovascular disease during their investigation. However, upon further review the study *only* evaluated 667 of those subjects who were diagnosed with cardiovascular disease during the study period. Therefore, about 39,000 surveys were eliminated. This design was perhaps the only way a study could not notice the benefit of vitamin C intake.

However, another interesting result sprouted from the Rimm study — the great value of vitamin E for heart patients. (In our view, vitamin E is especially beneficial for people already low in vitamin C.)

It is incredible that more than two decades after Pauling and Rath first published their *Unified Theory*, modern medicine and its schools, pharmaceutical companies, and the United States Government have all failed to make the slightest effort to investigate the effects of large amounts of vitamin C and the amino acid lysine on heart disease. More on this failure follows in The Pauling Therapy Handbook, Volume II.

This lack of interest is even more surprising when one considers that there are no proven treatments for atherosclerosis. According to a leading alternative medical doctor, coronary bypass surgery and angioplasty were never clinically "proven" before being adopted by the medical profession.

> A good example is angioplasty, in which a balloon on the tip of a catheter is used to open blockages. In my opinion, there is never a reason for anyone to have an angioplasty. It is a dangerous procedure looking desperately for validation. Whenever it is compared to a non surgical therapy - and there have been very few of these studies - patients treated with angioplasty virtually always fare worse. There is a higher death rate, higher heart attack rate and, in general, a higher repeat surgery rate. This procedure will, in my opinion, always be an unproven, expensive and dangerous gimmick that became an accepted therapy based on self-serving "presumption" only. Bypass surgery may be helpful for some patients, but it should not be used as the first treatment, and clearly not in mild heart attack patients. Medication, dietary and lifestyle changes, plus nutritional supplements are more effective approaches. — **Julian Whitaker, M. D.**

The neglected vitamin C theory is a unifying theory that encompasses homocysteine, lipid imbalances including Lp(a) excesses, infections, stresses, diet, free radicals, lesions, and a

71

lack of occlusive cardiovascular disease in most animals, and raises the important issue of mechanical stress via the bloodstream. The arguments are straight-forward, and most people can understand them. *If more cardiovascular patients could learn about the vitamin C theory, much suffering would end.* Pauling's recommended high-dose treatment has seen spectacular success among the members of the lay public who have discovered it.

Jeff Fenlason - After Ten Years, His Two-Day "Miracle" Recovery

Jeff Fenlason, formerly of North Carolina, was 52 years of age at the time of the reporting of this case. He is but one of thousands who have experienced the Pauling therapy miracle. Jeff is still living at the time of this writing. Jeff's case provides clear and convincing evidence in favor of the vitamin C theory. He has allowed his real name to be used, and much of his testimony is posted in Chapter 8.

Fenlason is important, in that he did not have EDTA chelation or any other alternative therapy before adopting the high-dose Pauling therapy. Jeff, as a military veteran, was entitled to complete medical care from the Veterans Administration and his decade of heart care was fully paid for. There was no reason to try alternative treatments. After ten years under the care of modern cardiology, Jeff's "two-day" reversal followed immediately after he began the Pauling therapy.

Alternative doctors who routinely offer their patients nutritional support are not likely to witness a Fenlason-style miracle. However, we witness this "miracle" almost every time a patient under the care of a modern cardiologist begins the

Pauling therapy, probably because these people were routinely told that there is no value in taking extra vitamin C.

A rapid "cure" requires 5 to 6 grams of the amino acid l-lysine with at least the same amount of vitamin C daily. (Note: Fenlason reports that he adopted a regimen of 14 grams of vitamin C daily, along with 5 to 6 grams of lysine.)

Just the Facts, Please

So how do scientists or members of the lay public select among the competing theories? The first step is to analyze data and try to identify facts. For purposes of this discussion we focus on cardiovascular disease, and not necessarily myocardial infarction (heart attack). The Pauling/Rath vitamin C theory is for occlusive cardiovascular disease — the how and why the plaques of atherosclerosis form mostly in the coronary arteries. (*It is true that arteries lined with plaque are not able to dilate in response to a clot as are healthy arteries, making a heart attack more likely.*)

Let's review the facts about cardiovascular disease, most of which have already been presented:

1. **The heart disease process begins with a lesion in the artery.** (Source: Brown-Goldstein Scientific American 1982, and Linus Pauling Heart Disease Video)

2. **Plaques are often localized to areas around the heart (coronary arteries) and the carotid arteries but do not occur within the heart itself.** (Source: An Experimental Study of the Intimal Ground Substance in Atherosclerosis, G. C. Willis, Canad M.A.J., July 1953, Vol 69, p 17, and *Stop America's #1 Killer*, Levy 2006)

3. **Veins, in general, do not suffer atherosclerotic plaque deposits.** (Source: common knowledge from medical literature)

4. **CVD leading to heart attack is the leading cause of mortality in the United States.** (Source: American Heart Association, http://www.amhrt. org/catalog/ Scientific_catpage70.html)

5. **In the United States the mortality <u>rate</u> from CVD peaked in 1970, having increased from a nominal rate at the beginning of the 20th century to its peak in the 1960s/1970s.** (Source: MMWR, February 16, 2001/50(06); 90-3, Mortality From Coronary Heart Disease and Acute Myocardial Infarction, United States, 1998, www.cdc.gov/ mmwR/preview/ mmwrhtml/mm4830a1.htm)

6. **Since 1970 the CVD mortality rate has declined by 40 percent from its peak (circa 1970) in the United States.** (Source: According to Linus Pauling in his 1986 book, *How to Live Longer and Feel Better,* the number of heart-related deaths in 1970 was around 740,000. The American Heart Association places the number of deaths in the year 2000 at between 400,000 and 500,000. Note the larger population in 2000.)

7. **After Pauling's first book on Vitamin C was published in 1970, vitamin C consumption increased by 300 percent in the United States.** (Source: Linus Pauling Institute of Science and Medicine)

8. **The United States is the only industrialized country to have experienced a 40-percent**

reduction in heart disease mortality. (Source: Linus Pauling Institute of Science and Medicine)

9. **Most high order mammals: (a) make their own vitamin C in high amounts (3 to 11 grams daily) in the liver or kidneys and, (b) do not suffer the same type of cardiovascular disease as humans.** (Guinea pigs do not make vitamin C and do suffer the same type of cardiovascular disease as humans when vitamin C is restricted in their diet.) (Various sources including Linus Pauling, *How to Live Longer and Feel Better*, 1986, and the Pauling/Rath *Unified Theory*)

10. **The Enstrom analysis (1992) of NIH data showed that vitamin C supplements of only 500 mg decreased the age-adjusted death rate in men by 40 percent (extending life by almost 6 years) and showed a somewhat lesser improvement in women.** (Source: Enstrom, Epidemiology 1992. Note: The Harvard/Rimm study obviously designed to "debunk" Enstrom only evaluated some 667 who developed CVD and ignored the dietary intakes of nearly 40,000 who did not get CVD.)

11. **Repeatable experiments with guinea pigs have shown that depriving the animals of vitamin C causes atherosclerosis that is quite similar to human lesions.** No plaque formed in the control group receiving "adequate" vitamin C. (Source: *The Reversibility of Atherosclerosis*, G. C. Willis, Canad M.A.J., July 15, 1957, Vol 77)

12. **Pauling and Rath showed that apo(a)/Lp(a) increased in the animals deprived of vitamin C**

but not in the controls. (Source: Immunological evidence for the accumulation of Lp(a) in the atherosclerotic lesion of the hypoascorbemic guinea pig, Pauling/Rath, Pro. Nat. Acad. Sci USA, Vol 87, pp 9388-9390, Dec 1990, Biochem)

13. **Vitamin C is required (and used up) in the making of collagen, the most abundant protein in the human body.** (Source: Roger J. Williams, Nutrition Against Disease, 1971, Linus Pauling, *How to Live Longer and Feel Better*, 1986)

14. **Scurvy manifests as a collagen disorder, which is caused by a lack of vitamin C in the diet.** (Source: James Lind, 1753, Linus Pauling, *How to Live Longer and Feel Better*, 1986)

15. **Lp(a), not ordinary LDL cholesterol, has the lysine and/or proline binding sites.** (Linus Pauling plus hundreds of Lp(a) papers in MEDLINE)

16. **The Beisiegel studies of post mortem human aortas in 1989 determined that plaque consists of Lp(a) only — no ordinary LDL cholesterol.** (Source: Morphological detection and quantification of lipoprotein(a) deposition in atheromatous lesions of human aorta and coronary arteries in Virchows Arch A Pathol Anat Histopathol 1990;417(2):105-11, Niendorf A; Dietel M; Beisiegel U; Arps H; Peters S, Wolf K; Rath M and Lipoprotein(a) in the arterial wall. Beisiegel U; Rath M; Reblin T; Wolf K; Niendorf A, Eur Heart J 1990 Aug;11 Suppl E:174-83)

17. **Elevated cholesterol has been correlated with CVD, but many studies were conducted before**

Lp(a) was known. The Lp(a) molecules were lumped in with LDL. In September 2000, an Oxford meta-analysis of 27 large studies showed that people with elevated Lp(a) are 70 percent more likely to suffer a heart attack or stroke. (Source: *Circulation* Sep 2000)

18. **Homocysteine levels rise when vitamin C levels are low.** Dr. McCully is cited in the Pauling/Rath *Unified Theory* paper because of his experiment that showed homocysteine levels to rise in vitamin C-deficient guinea pigs and not in controls. (Source: McCully KS, Homocysteine metabolism in scurvy, growth and arteriosclerosis. Nature 1971;231:391-392 from Pauling/Rath *Unified Theory*. See http://www.orthomed.org)

19. **Elevated homocysteine is considered a leading risk factor for (has been correlated with) heart disease.** (Source: Life Extension Foundation Treatment Protocols, Fibrinogen and Cardiovascular Disease, See http://www.lef.org/protocols/prtcls-txt/t-prtc149a.html)

20. **According to the Life Extension Foundation, 50 percent of all individuals 50 years or younger who die from heart disease succumb without any established risk factors.** (Source: Life Extension Foundation Treatment Protocols, Fibrinogen and Cardiovascular Disease, See http://www.lef.org/protocols/prtcls-txt/t-prtc149a.html)

21. **Primate experiments have indicated that a vitamin B6 deficiency in these animals causes**

atherosclerosis. (Source: Roger J. Williams, *Nutrition Against Disease*, 1971)

22. **Thomas E. Levy reduced all 27 known CVD risk factors to a single cause: a localized vitamin C deficiency.** This Board-certified cardiologist analyzed and found 650 studies to support his analysis. (Source: Thomas E. Levy, *Stop America's #1 Killer*, 2006)

There are common observations that may be relevant. The cardiovascular disease process seems to be accelerated in older people; in people who consume polyunsaturated fats without adequate vitamin E; and in people who undergo common coronary artery bypass surgery or angioplasty.

Popular writers during the 1950s, 1960s, 1970s, and later (e.g. Adell Davis) have helped disseminate the scientific knowledge of the health value of supplemental nutrition, especially vitamin C. As the knowledge became more widespread, the CVD mortality rate stopped increasing and began to decline.

Young people are also known to have CVD, and people as young as 30 can suffer heart attacks. Older people do suffer peripheral arterial disease (PAD), but CVD is generally localized to the carotid arteries and coronary arteries near or on the surface of the heart.

A frank copper deficiency has also been implicated in atherosclerosis.

I am not aware of any person who has consistently consumed more than 10,000 mg of vitamin C daily for several years who has then contracted cardiovascular disease. For those with prior disease, this amount usually controls any noticeable symptom of cardiovascular disease.

However, we are all individuals. Take, for example, Les, whose case is reported in Chapter 8. A few people require and can metabolize extraordinarily high amounts of oral vitamin C.

Given The Facts, Which Theory Fits?

The two crucial observations or facts that may help to sort through the several theories are: (a) animals do not generally suffer the same type of CVD as humans, and (b) plaque formations are uniform (not random) and usually occur only in arteries near the heart - *but not inside the heart.*

The first observation, that animals do not generally suffer CVD, indicates that there is a difference in the genetic makeup between humans and most animals. We fail to see how any popular theory, save perhaps diet, can explain this phenomenon.

The relevant genetic difference between animals and humans is our inability to make vitamin C. Other mammals make 3,000 to 11,000 mg of vitamin C in the liver (adjusted for body weight), which directly enters the bloodstream. Most humans cannot make a single molecule of vitamin C, and half of what we eat is not biologically available.

The second observation, that plaque deposits are uniformly close to the heart, tends to rule out poisons circulating in the blood (e.g. oxidized cholesterol, microbes, fats or homocysteine) as being the primary cause of CVD (although these factors probably aggravate or accelerate the disease). The cholesterol, homocysteine, oxidized cholesterol, microbe, fat, and even heavy metal/free radical theorists have yet to explain why plaque deposits are not randomly distributed throughout the bloodstream or why these blockages occur where the blood pressures and blood velocities are the highest.

Dr. Rath has pointed out that infarctions do not occur in the ears, nose or fingers, though they should if a poison injures an artery. We would expect more lesions away from the heart where the blood pools and travels more slowly. Likewise, something in the fast-moving blood, e.g. cholesterol, fat, homocysteine, or even Lp(a), cannot explain the coronary lesions or explain why these lesions do not occur in the coronary arteries inside the heart itself.

The vitamin C theory explains these observations as the body counteracting the detrimental effects of mechanical stress on arteries weakened by the lack of ascorbate. A low grade vitamin C deficiency may not kill us directly, but the deficiency leads to a suboptimal supply of collagen. The lack of collagen is blamed for the disruption of the ground substance surrounding the heart and its arteries. The resulting weakness leads to the lesions or cracks that develop and then attract Lp(a). The sticky Lp(a) is the reason cholesterol and other lipids become involved in the plaques. On the other hand, when the supply of collagen is adequate, as in most species that make their own ascorbate, blood vessels remain strong and resilient.

The vitamin C (collagen/Lp(a)) theory of atherosclerosis explains the vitamin B6 connection (also required to produce collagen) and copper (also required to produce collagen). Usually the deficiency of vitamin C is the primary factor. For example, vitamin B6 and copper are not usually implicated in scurvy.

Vitamin C's antioxidant properties and ability to support the immune system are well known, which may also explain away the oxidized cholesterol findings as well as the microbe theory.

Finally, McCully has shown that homocysteine levels are elevated in the scorbutic guinea pig and perhaps these levels even initiate atherosclerosis in the absence of vitamin C.

Conclusion

The common theories for heart disease fail in the face of two simple observations: Plaques do not generally form in animals that make their own vitamin C, and plaques do not form randomly in humans throughout the bloodstream. These two observations are the cornerstones of the Pauling/Rath vitamin C *Unified Theory*. The vitamin C *Unified Theory* holds that the LDL variant Lp(a) acts as a surrogate for vitamin C, and this theory is the only theory that also fits all other known facts and most observations.

That a vitamin deficiency causes heart disease defies belief, though the medical establishments' dismissive attitude toward vitamin C is precisely the reason Pauling himself gave for writing his first vitamin C book. According to the book, *Linus Pauling In His Own Words*, edited by Barbara Marinachi, Pauling was appalled by the medical profession's bias against vitamins, especially ascorbic acid.

The author welcomes additional facts and observations but has become convinced, based in large part on the nearly universal positive anecdotal reports of clinical response received, that the Pauling/Rath theory is correct and that over time the Pauling high vitamin C/lysine therapy can actually reverse the lesions characteristic of cardiovascular disease.

The author does not understand why humans have evolved so differently from most animals. How have we survived as a species without the ability to manufacture our own vitamin C? One is reminded of the fictional "lysine contingency" in the movie *Jurassic Park*. (Perhaps He who made us in His image decided to incorporate a similar built-in control mechanism for humankind?)

Theological questions aside, there is now little doubt in the author's mind that this great deficiency in our genetic makeup, the lost ability to produce the liver enzyme L-gulonolactone oxidase (GLO), an enzyme that would otherwise allow us to convert ordinary glucose (sugar) into ascorbic acid, is what ultimately causes occlusive cardiovascular disease.

America's leading killer is a genetic disease.

"Knowing that lysyl residues are what cause Lp(a) to get stuck to the wall of the artery and form atherosclerotic plaques, any physical chemist would say at once that the thing to do is prevent that by putting the amino acid lysine in the blood to a greater extent than it is normally." — Linus Pauling, *Journal of Optimum Nutrition*, Aug 1994

Chapter 6

Lp(a) Binding Inhibitors

The liver produces more Lp(a) molecules in humans with chronically low intake of vitamin C. These numerous Lp(a) molecules have a tendency to deposit on top of existing plaque formations. Over time, the healing process overshoots and arteries narrow. The resulting reduction in blood flow introduces several dangers, including a greater likelihood that any blood clot will cut off the flow of blood entirely.

According to Linus Pauling, this problem has a solution. He announced in the year 1992, "We're at the point where I think we can get almost complete control of cardiovascular disease, heart attacks and strokes."

The key to a possible "nonsurgical" treatment for heart disease came when highly regarded biochemists and chemists uncovered the direct means by which atherosclerotic plaques begin to form. Brown and Goldstein's discovery that this process begins when Lp(a) is attracted to amino acids within the lesion is of great significance. As the walls of arteries break down, strands of lysine (and proline) that were once a part of the collagen helix become exposed. Pauling called these exposed strands "lysyl and prolyl" residues.

The great chemist explained, "You need lysine to be alive, it is essential, you have to get about one gram a day to keep in protein balance, but you can take lysine, pure lysine, a perfectly non toxic substance in food as supplements, and that puts extra lysine molecules in the blood. They enter into competition with the lysyl residues on the wall of arteries and accordingly count to prevent Lp(a) from being deposited or even will work to pull it loose and destroy atherosclerotic plaques."

Lysine is one of twenty essential amino acids. Amino acids are the building blocks of protein. During digestion proteins are broken down into their constituent amino acids and absorbed into the bloodstream. The DNA in our cells contains the instructions for reassembling the amino acids into human proteins.

Amino acids come in two varieties — left and right. The left-handed version, l-lysine, is the form our bodies use. If you purchase lysine as a supplement, you do not have to worry that it will be l-lysine. Likewise, if you purchase vitamin C as ascorbic acid, it will be the only form with vitamin activity, i.e. l-ascorbic acid.

The word *essential* implies that we require lysine in the diet just as we require vitamins and trace minerals. The word *nonessential* is misleading. We require nonessential amino acids as well but our bodies are able to manufacture the nonessential ones. Proline is an example of a nonessential amino acid that our cells can manufacture.

Reversing Heart Disease By Nullifying the Lysine Binding Sites

Each Lp(a) molecule has a finite number of lysine binding sites — points of attachment to lysine. The Pauling/Rath invention is the means to inactivate Lp(a) molecules, making them less "sticky" and thus preventing them from forming or growing the plaque buildups. The "cure," if you will, for heart disease is straightforward, inexpensive, and non-toxic. Heart patients need merely to increase their intake of lysine along with their increased intake of vitamin C.

As more lysine enters the bloodstream, the probability increases that Lp(a) molecules will find plaques less attractive. If the Lp(a) binds with free lysine floating in the blood, it cannot later bind with the lysine (or proline) that has been exposed on the atherosclerotic plaques. As the lysine *keys* attach to the Lp(a) *locks*, the keyholes called *receptors* become filled. More lysine in the blood therefore reduces the sticky property of Lp(a) required for making "plaster casts."

According to Linus Pauling, extra lysine in the blood will even reverse atherosclerosis. "It works to pull Lp(a) loose and destroy atherosclerotic plaques," said Pauling.

After the binding receptors of all of the Lp(a) molecules are filled with the free lysine floating in the blood, the Lp(a) molecules become as harmless as ordinary low density lipoprotein (LDL) cholesterol. Ordinary LDL cholesterol is not only harmless but essential for life and good health (see Chapter 10).

Pauling and Rath called the nutritional substances that can treat chronic scurvy and destroy existing plaques *Lp(a) binding inhibitors*. These scientists have been awarded three United States patents for binding inhibitor formulas that destroy atherosclerotic plaques *in vitro* and *in vivo*. These patents establish their preeminent position and should keep others in orthodox medicine from someday claiming credit. The mechanism of action of Lp(a) binding inhibitors is best described in their first U. S. patent 5,278,189, granted on January 11, 1994.

The primary binding inhibitors are vitamin C to increase collagen production, which in turn improves the health and strength of the arteries, and lysine to promote collagen, nullify Lp(a), and dissolve plaques. These substances taken together are clinically effective. Neither lysine nor vitamin C has any known toxicity or lethal dose.

Lp(a) binding inhibitors become the *Pauling therapy* for heart disease only at sufficiently high dosages, i.e. between 3 and 18 grams of vitamin C (ascorbic acid) and 3 and 6 grams of lysine.

Today, nearly 20 years of consistent testimony demonstrates that Pauling's recommended dosages of the Lp(a) binding inhibitors are almost always effective for reversing symptoms of advanced heart disease within 30 to 90 days.

The major discovery that human Lp(a) binds to lysine preceded the discovery that Lp(a) also has proline binding receptors, which means Lp(a) will also bind with the amino acid proline. Subsequently, proline was found to be an even more effective Lp(a) binding inhibitor than lysine *in vitro*. Alternative doctors now recommend the addition of between .5 and 2 grams of proline to the Pauling therapy, as this may be of significant additional benefit to heart patients.

Pauling's Unified Theory Lecture

In his *Unified Theory* lecture, Linus Pauling recounted how he invented the lysine treatment, and then discussed the first three known cases where high doses of vitamin C and l-lysine worked "miraculously." Linus Pauling published these first few case studies in the *Journal of Orthomolecular Nutrition*. Unfortunately, this Journal is not catalogued by the National Institutes of Health's online MEDLINE database. (*Medical doctors have never seen these cases because Pauling and Rath, as have most researchers in alternative medicine, found it impossible to get their work published in mainstream medical journals. Even today, MEDLINE does not include many alternative medical journals.)*

Abram Hoffer, M.D., Ph.D., is now well known in the alternative and orthomolecular medical communities. He was the first to conduct and then publish a double-blind, randomized control trial. Dr. Hoffer's work also heightened Dr. Pauling's interest in the efficacy of vitamins at doses much higher than those required to cure frank vitamin deficiencies. Dr. Hoffer expressed to me in the following correspondence his concern with the current restrictions placed on the MEDLINE database:

Dear Mr. Fonorow:

Your report on Chronic Scurvy in Nexus New Times Jan-Feb 2006 is very good and important. You are properly concerned with the lack of attention given to Dr Pauling's excellent work. One of the reasons is that his basic clinical work has been ignored by MEDLINE as he published it first in our journal, the Journal of Orthomolecular Medicine. Even the Proceedings of the National Academy of Sciences Washington refused to publish his clinical material. This is based on the foolish view that he did not have an M.D. The fact that he had nearly 40 Ph.D. and D.Sc.'s and other honors was ignored. Therefore, as MEDLINE has consistently refused to abstract our Journal, his original reports are relatively unknown. I consider this policy of MEDLINE unforgivable and have been for the past 35 years trying to change their minds. The debate has been taken up by Andrew Saul and Dr. S. Hickey, and they are both doing a fantastic job of creating pressure against MEDLINE policy.

As an example a scientist at a U. S. University found that there was a problem with schizophrenic brains in transforming tryptophan into NAD. I had originally suggested this many years ago but this work remained unknown, and when I pointed this out to the scientist she apologized but it was not her fault. It was the fault of MEDLINE, which has deliberately done its best to keep this information from world scientists. This must come to an end.

It occurs to me that you and your Web site might look into this, because it does explain why his work is not better known. And it is not entirely due to the ill will and stupidity of the establishment. They are ignorant, too, and MEDLINE helps keep them ignorant.

Abram Hoffer

Before/After Surgery Demonstrates Reversal

At first I, too, was skeptical and not totally convinced that Pauling and Rath had uncovered the root cause of the great problem of cardiovascular disease.

In 1996 my aunt called me after my uncle was involved in an automobile accident. She told me that my uncle had been through his second operation but that this time his doctors could not find the white plaque that they had seen in his first operation where an ultrasound had predicted it should be. Here is the story.

Following the car accident a routine physical examination revealed that my uncle had blockages in both of his large neck (carotid) arteries. Based on his ultrasound it was estimated that he had a 90-percent blockage on one side and a 50-percent to 60-percent blockage on the other. Two surgeries (carotid endarterectomy) were scheduled by my uncle's longtime family doctor to improve blood flow by scraping the white atherosclerotic plaque from the inside of my uncle's arteries.

The procedure was performed on the artery with the 90-percent blockage in May of 1996 and the white plaque was removed. (The doctor showed my aunt and uncle the plaque he had scraped from my uncle's occluded artery.) A second procedure to address the other carotid artery was scheduled for one month later in June.

My uncle began to feel much better after the first surgery but my aunt decided that she would take no chances. She gave my uncle vitamin C and lysine. My uncle did not believe in taking vitamins and resisted taking the supplements, though my aunt insisted. After the first surgery she had learned what I was involved in from talking to her sister (my mother). My aunt was able to convince my uncle to follow Pauling's recommendations, i.e. to take both vitamin C and lysine. Both nutrients are available from any drug store.

I subsequently learned that my aunt had made a mistake in the dosage. She had cut Pauling's recommended dosage in half. Instead of giving my uncle 5 grams of each (or 5,000 mg), she gave him 5 tablets of each spread throughout the day. The tablets were 500 mg each, so the daily dosage my uncle received for that month was 2,500 mg of vitamin C and 2,500 mg of lysine. My aunt took all manner of pains to make sure that my uncle consumed the vitamin C and lysine every day for that month. She told me that this was the only change she made in his regimen.

The second carotid endarterectomy for the 50-percent to 60-percent blockage was performed one month later in June 1996. My uncle's doctors were astounded because this time they could not show my aunt and uncle any of the plaque such as they had scraped with the first procedure. There was no blockage where the ultrasound had predicted it to be!

During the excited telephone call after the second surgery, my aunt told me that she couldn't help but notice the bruises from the second procedure. The first surgery had not produced any bruising. She had a different report after the second surgery: "Your uncle is bruised up and down his neck." Apparently the doctors were looking for but couldn't find the blockage that they knew just had to be there.

My aunt and uncle never mentioned the vitamin C/lysine therapy to his doctor, but we have since written to him and requested copies of the relevant medical records, along with permission to use them on the Internet. To date, the doctor has not responded, perhaps partly in fear of a malpractice lawsuit for performing the "unnecessary" second surgery.

This case is important because it establishes that following Pauling's basic advice of taking vitamin C and lysine by themselves at even half the recommended dosage can reverse an estimated 50-percent blockage in a large artery within one month. I published this case report in the 1996 *Life Extension* magazine.

Linus Pauling's one-hour lecture is now available on DVD video from the Life Extension Foundation. In his lecture Pauling recounts the first cases where his high vitamin C and lysine therapy quickly resolved advanced cardiovascular disease in humans. The effects were so pronounced and the inhibitors so non-toxic that Pauling mentioned in his lecture that he doubted a clinical study was even necessary before advising the public.

Linus Pauling was the first to report the highly beneficial effects of his invention, although he prudently recommended a gradual approach. Even with a gradual intake of vitamin C and lysine, the good results were noticeable in periods of as little as two to four weeks. There is now an abundance of reports, many by people willing to use their actual names, that attest to the pronounced effects of this treatment, especially in patients with severe disease. Samples of these reports are presented in Chapter 8.

When Pauling was asked whether he really thought this development represented the cure for heart disease, he responded:

> I think so. Yes. Now I've got to the point where I think we can get almost complete control of cardiovascular disease, heart attacks and strokes by the proper use of vitamin C and Lysine. It can prevent cardiovascular disease and even cure it. If you are at risk of heart disease, or if there is a history of heart disease in your family, if your father or other members of the family died of a heart attack or stroke or whatever, or if you have a mild heart attack yourself, then you had better be taking vitamin C and lysine." — **Linus Pauling Interview, JOM 1994**

Treating Elevated Lp(a)

When serum Lp(a) is elevated above the normal range, Lp(a) binding inhibitors can profoundly interfere with the disease process. Keeping in mind that elevated Lp(a) is the body's response to chronic scurvy, vitamin C to tolerance dosage is always indicated.

Anecdotally, binding inhibitor formulas that include proline have been documented to lower Lp(a) in 6 to 14 months. In cases where Lp(a) is not reduced, binding inhibitors become even more important in the fight against plaques, regardless of their effect on lowering serum Lp(a).

Proline

Linus Pauling had limited his recommendations to vitamin C and lysine, and I witnessed strong, first-hand evidence (in my uncle) that Pauling's simple, non-toxic, and inexpensive combination works wonders. However, Pauling's close associate, Dr. Matthias Rath, the first Director of Cardiovascular Research at the Linus Pauling Institute, made a somewhat different recommendation. Rath expanded the recommendation to include another amino acid, proline, and generally recommended lower dosages of these two amino acids, judging by his products.

Former Vice President Al Gore worked to allow the general public to gain free access to the MEDLINE database on the National Institutes of Health's World Wide Web site. For this he should be commended.

I began to explore MEDLINE when searching for the reason that Pauling and Rath had made differing recommendations. My curiosity regarding this difference in therapeutic agents and dosage recommendations began a fruitful relationship with the "free" MEDLINE Internet search engine, Internet Grateful Med.

There has been intensive research establishing that it is Lp(a) and *not* LDL cholesterol that exhibits the Lysine Binding Site (LBS). Thus, it is Lp(a) that adheres or binds to lysine residues in damaged arteries and leads to plaques. The role of Lp(a) is now well understood.

The rhesus monkey, like humans, cannot make its own vitamin C. The Lp(a) found in rhesus monkeys, however, does not bind to lysine and has thus been designated as LBS-. The rhesus monkey's Lp(a) does have a "proline binding domain," and thus this peculiar form of Lp(a), LBS-, binds to fibrinogen.

Furthermore, researchers have been able to show that proline in the test tube may interfere with the so-called "assembly" or creation of Lp(a). Researchers at the University of Chicago were the first to demonstrate this "proline binding region," and in doing so they have shown that proline and a lysine analog can inhibit the binding of Lp(a) in vitro, i.e. in the test tube.

I found the following MEDLINE abstract, which prompted me to contact these researchers:

It is now established that the lysine binding site (LBS) of apo(a) kringle IV-10, and particularly Trp72, plays a dominant role in the binding of lipoprotein(a) [Lp(a)] to lysine. To determine the role of the LBS in the binding of Lp(a) to fibrinogen, we examined the binding to plasmin-modified (PM) fibrinogen of human and rhesus monkey Lp(a) species classified as either Lys+ or Lys- based on their capacity to bind lysine Sepharose and to have Trp or Arg, respectively, in position 72 of the LBS of kringle IV-10. We also examined the free apo(a)s obtained by subjecting their corresponding parent Lp(a)s to a mild reductive procedure developed in our laboratory.

Our results show that both Lys+ and Lys- Lp(a)s and their derived apo(a)s, bound to PM-fibrinogen with similar affinities (Kds: 33-100 nM), whereas the B(max) values were threefold higher for apo(a)s. Both the lysine analog epsilon-aminocaproic acid and L-proline inhibited the binding of Lp(a) and apo(a) to PM fibrinogen. We conclude that the LBS of kringle IV-10 is not involved in this process and that apo(a) binds to PM-fibrinogen via a lysine-proline-sensitive domain located outside the LBS and largely masked by the interaction of apo(a) with apoB100. The significant difference in the PM fibrinogen binding capacity also suggests that apo(a) may have a comparatively higher athero-thrombogenic potential than parent Lp(a). — **Department of Medicine, University of Chicago, Illinois 60637, USA**

Dr. Olga Klezovitch, Ph.D. was kind enough to respond to my letter:

Dear Owen Fonorow:

Our study published in Journal of Clinical Investigation demonstrated for the first time that L-proline can inhibit the binding of Lp(a) and apo(a) to fibrinogen in in vitro conditions. However, we have no data about the physiological relevance of this finding. It is well established that the binding of Lp(a) to fibrinogen also involves the lysine binding sites of apo(a) (component of Lp(a)), therefore, the presence of lysine and lysine analog dramatically reduces this binding. Once again, the in vivo relevance of this inhibition is unknown.

Sincerely,

Olga Klezovitch, Ph.D.
University of Chicago
Department of Medicine

Based on this research, alternative medical doctors familiar with the therapy, and the companies producing Pauling therapy products, now add the amino acid proline to the Pauling protocol. This nutritious addition seems to make Lp(a) binding inhibitors stronger and work faster.

One caveat is that proline is not essential, meaning that cells in the human body make proline. In all probability our bodies make less proline as we age, leading to the higher prevalence of CVD in the elderly. The problem is one of determining proper dosage. The quantity of proline an individual makes endogenously is unknown. This may be one reason Pauling did not specifically recommend proline.

Why the Pauling Therapy is Not a "Cure" for Heart Disease

Technically, the high vitamin C/lysine Pauling therapy is not the "cure" for heart disease because it does not overcome the genetic defect in the human liver. Heart disease is apparently "cured" in most people, so long as they follow Pauling's recommendations. However, patients should not stop taking vitamin C in large amounts or the condition will eventually reappear.

Unlike animals, we humans cannot make our own vitamin C and must obtain the "right" amount every day. The right amount is certainly larger than we can get through a normal diet. (I estimate the right amount of vitamin C is between 6,000 mg and 18,000 mg of ascorbic acid daily. Over the past 12 years I have never encountered heart disease in any person who takes more than 10,000 mg of vitamin C daily.)

Pauling used the more precise terminology when he said "we can *completely control* heart disease" with vitamin C and lysine.

There are some guinea pigs that can produce their own vitamin C and there are a few humans who probably possess this ability as well. For the vast majority, however, the cure will only become a reality when genetic engineering corrects the GLO defect in our DNA. When corrected, the human liver would produce its own ascorbate out of the sugar glucose.

"Although physicians, as part of their training, are taught that the dosage of a drug that is prescribed for the patient must be very carefully determined and controlled, they seem to have difficulty in remembering that the same principle applies to the vitamins." — Linus Pauling

Chapter 7

The Pauling Therapy

Cardiologists have been kept in the dark regarding the vitamin C connection to heart disease. Most cardiovascular drugs are compensating for low vitamin C intake. There are cardiovascular drugs that exacerbate heart conditions. In my opinion, for the best patient response the doctor would be well advised to replace as many standard heart medications as possible with the following vitamin C and lysine protocol.

NOTE: Linus Pauling specifically recommended high, generally equal oral doses of vitamin C and the amino acid lysine between 5,000 and 6,000 mg in his Unified Theory lecture (available on video). Anything less, by definition, is not the Linus Pauling Therapy.

The extended protocol includes advice given by Linus Pauling in his 1986 book, *How To Live Longer and Feel Better.* The other recommendations account for variables such as a poor diet, advancing age, and/or the use of the prescription drugs commonly given to heart patients. I have attempted to present the additional nutritional substances in the order of their importance.

Linus Pauling coined the term *orthomolecular* — right molecules — to stand for vitamins, amino acids, and other non-toxic molecules which our bodies are familiar with. Orthomolecular nutrients are generally devoid of toxicity and can be safely taken in much larger amounts than *toximolecular* prescription drugs.

Note that trace minerals are only orthomolecular in trace amounts and, unlike vitamins, may become toxic at higher dosages.

The Basic Orthomolecular Recommendations for Controlling Heart Disease

Linus Pauling Recommendations

1. Take 6,000 to 18,000 mg of vitamin C as ascorbic acid daily (*some of the vitamin may be taken as sodium ascorbate*) **up to bowel tolerance (6 to 18 g).**

Pauling's therapeutic dosage of vitamin C for those diagnosed with cardiovascular disease is from 6,000 mg up to 18,000 mg (or bowel tolerance). Generally, 3,000 to 6,000 mg of vitamin C is the recommended preventive dosage.

The half-life of vitamin C in the bloodstream is 30 minutes. Linus Pauling advised taking vitamin C throughout the day in divided doses. The Hickey/Roberts Dynamic Flow theory predicts that taking vitamin C every four hours will produce the highest sustained blood concentrations. Take more before bedtime.

Those who have bloating, gas, or diarrhea after taking vitamin C should reduce the dosage of ordinary vitamin C and consider adding a liposomal form of vitamin C such as the *Lypo-Spheric*™ vitamin C product available from livonlabs.com

2. Take 2,000 to 6,000 mg (2 to 6 g) of lysine daily.

For those diagnosed with cardiovascular disease, Linus Pauling recommended taking 5,000 to 6,000 mg of lysine daily. He recommended supplementing with at least 2,000 mg daily for prevention.

The following excerpt is from the *Unified Theory* lecture. Linus Pauling relates the story of his invention of the Pauling therapy for cardiovascular disease, which was to add lysine to vitamin C. Dr. Pauling explains what happened in the case of the first person to try the therapy, a distinguished anonymous scientist who had asked Pauling for advice. The scientist was on disability, in pain, and generally unable to do work or exercise despite taking 5,000 mg of vitamin C daily. He asked Linus what else he might recommend for his cardiovascular disease, and Pauling recounts his own response as follows:

I didn't have to tell him that lysine is an essential amino acid and you have to get around a gram a day to be in good health, and you get it in your foods, because he is one of the most distinguished biochemists in the United States, recipient of the National Medal of Science in the United States. So he said, "How much shall I take?" I thought, "What do I know?" I know that people get a gram or two in their food depending upon how much meat and fish they eat, that it's essential, that they have to get around one gram. It hasn't any known toxicity in animals or human beings. I said, "5 grams, 5 grams of lysine per day." He thanked me.

A couple of months later he telephoned me and said, "It's almost miraculous! I started taking a gram a day and 2 grams and so on. Within a month after I had reached 5 grams a day of lysine in addition to my 5 grams of vitamin C, I could walk two miles without any nitroglycerin tablets or without any pain in the chest." He said he had cut down the amount of heart medicine in half. "It's almost miraculous," he said.

Another couple of months went by and he telephoned me and said, "I was feeling so good the other day that I cut down a big tree in our yard and was chopping it up for wood, and I was also painting the house, and I got chest pains," this despite his 5 grams of vitamin C and lysine. So he said that he "went up to 6 grams of lysine and 6 grams of vitamin C and told me, "Now I am continuing chopping down, chopping up the tree and painting the house." And now a couple of years later he is still in fine health. — **Linus Pauling [Transcribed from his 1993 Linus Pauling *Unified Theory* Lecture]**

3. Follow Linus Pauling's general heart and cardiovascular nutritional recommendations as provided in his 1986 book, *How to Live Longer and Feel Better*.

In addition to 6,000 to 18,000 mg of vitamin C, Linus Pauling advised:

Vitamin E - 800 IU (to 3,200 IU)
Vitamin A - 20,000 to 40,000 IU
Super B-Complex - one or two
Daily multiple vitamin/mineral supplement
Eat less sugar
Drink plenty of water

Matthias Rath Recommendations

4. Take 250 to 2,000 mg of the amino acid <u>proline</u> daily.

This nutrient is an addition to the original Linus Pauling protocol. Dr. Rath specifically recommends the amino acid proline because of its strong Lp(a) binding inhibitor properties *in vitro* (in test tube experiments). There is anecdotal evidence that proline supplements lower elevated Lp(a) over a period of 6 to 14 months.

It is difficult to suggest an optimum dose for everyone because the healthy body manufactures its own proline, though we probably make less as we age. A few alternative doctors have recommended 2 g (2,000 mg), although the first Pauling therapy formula, Tower's *Heart Technology*™, has produced consistently good results with 400 mg per serving. Two servings daily (800 mg of proline) were shown to lower Lp(a) in a small pilot study.

Additional Enhancements

Although studies are lacking, there has been some research and considerable anecdotal experience since Pauling and Rath first published their theory. This knowledge has led us to make the following additional recommendations:

5. Take 100 to 300 mg of coenzyme Q10.

Coenzyme Q10 (CoQ10) is a vital substance required for the production of energy in all cells and plays an important role in maintaining proper heart function. Popular heart medications interfere with the body's own production of CoQ10; thus taking these drugs at high dosages may lead to heart failure. If you are always tired or your muscles ache, take more CoQ10. The human body makes CoQ10 but we lose this capability gradually as we age. Note: Vitamin C and several vitamins may help stimulate your own synthesis of CoQ10.

6. Eliminate trans fatty acids from the diet and introduce unprocessed Omega-3 and Omega-6 oils.

Medical doctors often call me after reading one of my articles to ask, "If Linus Pauling was correct in claiming that a chronic vitamin C deficiency causes cardiovascular disease, why then do patients respond so well after we put them on Omega-3 fish oils?" There was no definitive answer to this question until the writings of Mr. Thomas Smith of healingmatters.com provided one.

Mr. Smith is a lay person who was once a Type II diabetic. He reviewed the available scientific literature and used this knowledge to cure his own diabetes in three months. According to Smith's research, Type II diabetes is a disturbance of the cell membranes that blocks the uptake of glucose from the blood. Because vitamin C shares the same insulin-mediated transport through the membrane, vitamin C faces the same theoretical difficulty entering the cells of diabetics. The addition of Omega-3 fatty acids in place of trans fatty acids in the diet will rectify this disturbance of the cell membrane, according to Smith.

The Pauling therapy will be more effective if the heart patient reduces sugar and simple carbohydrates; eliminates man-made/processed fats such as trans fats and hydrogenated oils; and supplements Omega-3-rich oils such as evening primrose, flaxseed, and certain fish oils, because more vitamin C can enter cells.

"Research has shown that an Omega-3 Index of 8 to 10 percent reduces a person's relative risk of death from coronary heart disease by 40 percent, and from sudden cardiac death by 90 percent." This benefit probably results from restored insulin-mediated glucose/vitamin C uptake into cells.

Note: Following an Atkins-style, low-carbohydrate diet will eliminate most trans fats because these "poisons" appear mostly in processed carbohydrate foods such as cookies, crackers, snacks, etc. Butter is superior to margarine. Natural saturated fats are superior to any fats or oils processed for longer shelf life.

7. Eat salt, but only unrefined salt.

This advice might seem strange, but alternative medical doctor David Brownstein uncovered literature showing that a low-salt diet can cause the body to change its hormonal balance as it attempts to retain sodium. This imbalance leads to a 400-percent chance of heart attack in those with high blood pressure and low sodium intake. Refined (ordinary table salt) is probably poisonous, but <u>unrefined</u> salt such as *Celtic Sea Salt* contains over 80 minerals and can be considered a necessary "health food."

8. Supplement with magnesium (150 to 1,500 mg).

Certain chelated forms of magnesium are better absorbed and you can take less. The heart requires magnesium, and a deficiency should be corrected to maintain a regular heartbeat. Along these same lines, <u>supplemental manganese (Mn) should be reduced</u> (intake should be no more than 2 milligrams). More than 20 mg daily of manganese (Mn) can lead to irregular heartbeat, according to researchers at the United States Department of Agriculture (USDA).

I learned about the problem of manganese toxicity from a male caller who claimed that he had cured a life-long heart arrhythmia. Using the Internet to research his condition, the caller had found and contacted USDA researchers in South Dakota. The USDA scientist told him that the heart has an equal affinity for both manganese (Mn) and magnesium (Mg) but requires copious amounts of magnesium (Mg) for a regular heartbeat. Too much manganese (Mn) crowds out magnesium uptake and causes an irregular heartbeat.

The USDA scientist noted that 20 mg of manganese (Mn) daily will cause the problem and recommended reducing manganese to 2 mg daily. Furthermore, the researcher said that it can take 60 days for the body to detoxify and rid itself of excess manganese before the heartbeat becomes regular.

The caller found that by studying labels and adding up the manganese in his supplements he was indeed consuming more than 20 mg of manganese. He cut back to 2 milligrams daily and in 60 days, as the USDA researcher had predicted and for the first time that he could remember, his heartbeat became normal.

The caller contacted the USDA researcher in South Dakota to thank him, but for some reason the researcher denied having given him this advice. The caller then contacted me. During the investigation we noticed that the USDA papers that had been posted had been taken down. It seems likely that USDA researchers, like myself, are not supposed to offer medical advice to people or study the effects of nutrients on humans. After one or two years, several new papers appeared at the USDA Web site including a paper with the following abstract: *"Manganese alters mitochondrial integrity in the hearts of swine marginally deficient in magnesium....These results suggest that high Mn, when fed in combination with low Mg, disrupts mitochondrial ultrastructure and is associated with the sudden deaths previously reported"* (USDA Researchers).

9. Eliminate ordinary sugar and refined carbohydrates.

New research confirms Dr. John Ely's 30-year theory that sugar (glucose) competes with ascorbic acid (vitamin C) for insulin-mediated uptake into cells. Consuming sugar and refined carbohydrates effectively crowds out vitamin C and prevents it from entering cells, even in non-diabetics. Diabetics are especially vulnerable because their elevated blood sugar levels already reduce vitamin C's chances of entering their cells.

10. Supplement with vitamin K, either the K1 or K2 form (1 to 40 mg K1 or 150 mcg K2).

A function of vitamin K is to regulate calcium from soft tissues into bones. The prescription blood "thinning" drugs such as Coumadin (warfarin sodium) interfere with vitamin K. These drugs are derived from rat poison and have been proven to cause rapid calcification of soft tissues in animal studies. There is also new evidence that they cause hard arteries in humans.

Unfortunately, these drugs are routinely prescribed. Patients on "rat poison" style blood thinners should avoid vitamin K until they find a nutritionally-oriented physician to help wean them. Blood thinners that may substitute for these prescription drugs include: 2,000 IU of Unique-E (natural vitamin E from A. C. Grace); the amino acid arginine (3,000 mg); grape seed extract; and high dose Omega-3 fish oils.

11. Avoid supplemental calcium.

According to author Bill Sardi, Americans generally obtain the organic calcium they require in the diet. According to Dr. Levy, most supplemental calcium is of poor quality. If you elect to supplement with calcium, you should be sure to supplement with at least the same amount of magnesium.

12. Add a multivitamin/multi-mineral supplement as insurance.

13. Supplement with the amino acids taurine, arginine and carnitine (1 to 3 g).

All three of these amino acids (or amino acid-like nutrients) in higher dosages benefit muscle and heart function.

14. Add supplemental vitamin D3, especially in the winter months (2,000 IU).

Pauling did not recommend supplementing vitamin D partly because of a known toxicity of the synthetic D2 form and partly because sunlight creates vitamin D from cholesterol beneath the skin. However, our skin production of vitamin D is limited to periods of sunlight exposure, and in some latitudes, e.g. north of Atlanta, Georgia, vitamin D is only generated during the summer. Even during the summer months the UV/B light in the sunlight can only stimulate the production of vitamin D between 10:00 a.m. and 2:00 p.m. It is estimated that 20 minutes of sunbathing when the UV/B light can penetrate the atmosphere creates about 8,000 IU of vitamin D. The toxicity issue is avoided by supplementing the D3 form of the vitamin. In addition to its cardiovascular and anti-cancer properties, vitamin D3 also provides significant protection against cold and flu viruses.

15. Take 3 to 6 mg of melatonin before bed.

Melatonin is the sleep hormone and one of the most potent antioxidants known to science. A small gland in the brain known as the pineal produces melatonin in the absence of light. Many heart patients on prescription statin drugs have difficulty sleeping. If you are over 40 years of age and have difficulty sleeping, your pineal gland may not be making enough melatonin. Take 3 to 6 mg of melatonin just before bed to improve sleep. Melatonin researchers praise its wide-ranging longevity and cardiovascular and anti-cancer benefits. There are so many benefits attributed to melatonin that the list even exceeds the benefits of vitamin C. Coincidentally, the half-life of melatonin in the blood is 30 minutes. Melatonin, like alpha lipoic acid, is both fat and water soluble. This natural hormone has been found in every cell, both plant and animal, and has no known toxicity.

Pauling Therapy Summary

Therapeutic

Vitamin C (6,000 to 18,000 mg)
Lysine (5,000 to 6,000 mg)

Pauling Therapy Enhancements

Proline (250 to 2,000 mg)
Coenzyme Q10 (100 to 300 mg)
Magnesium (150 to 1,500 mg)

Preventives

Vitamin C (3,000 to 10,000 mg)
Lysine (2,000 to 4,000 mg)

Follow Pauling's Other Heart and Cardiovascular Recommendations

Vitamin E - 800 to 3,200 IU
Vitamin A - 20,000 to 40,000 IU
Super B-Complex - 1 or 2
Daily multiple vitamin and mineral
Drink plenty of water

Additional Enhancements

Eliminate trans fatty acids from the diet
Introduce unprocessed Omega-3 and Omega-6 oils
Eat salt, but only unrefined salt
Reduce manganese intake
Eliminate ordinary sugar and refined carbohydrates
Supplement with vitamin K
Avoid supplemental calcium
Supplement with the amino acids taurine, arginine and
    carnitine (1 to 3 grams)
Supplement with vitamin D3 (2,000 IU), especially in
    the winter months
Supplement with melatonin (3 to 6 mg) before bedtime

"I've got to the point where I think we can get almost complete control of cardiovascular disease, heart attacks and strokes by the proper use of vitamin C and lysine. It can prevent cardiovascular disease and even cure it. If you are at risk of heart disease, or if there is a history of heart disease in your family, if your father or other members of the family died of a heart attack or stroke or whatever, or if you have a mild heart attack yourself then you had better be taking vitamin C and lysine." — Linus Pauling 1994

Chapter 8

Important Case Studies

The beneficial effects of Pauling's therapy at the right dosages are pronounced, rapid, remarkable, and difficult to attribute to anything else. Initially our conversations were only with hopeless cases — people who were literally sent home to die. These were the people who, almost without exception, recovered on the therapy. The sickest patients reported experiencing the most relief. They had been told that one or more of their coronary or carotid arteries were 90 percent or more blocked, and that for one reason or another surgery was no longer an option. They had suffered heart attacks, coronary artery bypass surgeries, and angioplasties, and most complained of severe angina pain. Over the years we have often heard the refrain, *"My angina pain went away after 10 days"* from people who began the therapy.

Mr. Eli Raber contacted us in early 1997. He was in constant pain and was told by his doctors that he had run out of veins that could be used for another bypass surgery. His doctor had suggested that Eli search the Internet because there was nothing more that could be done. (The doctor may have been hoping that Mr. Raber would find an EDTA chelation alternative doctor. Instead, this man found our Web site.)

Six months later, Mr. Raber told Owen that his intractable pain had stopped 10 days after adopting Pauling's therapy, though he didn't relate that information at the time because he thought that his pain would recur. Mr. Raber wrote the following four years after starting the Pauling therapy:

February 2001

To whom it may concern:

My name is Eli Raber, Blackville, SC. Let me tell what the Pauling therapy is doing for me. In December 1996 I had a heart cath and was told that the arteries in my heart are severely blocked. One side is completely blocked, the other side is from 70 to 90% blocked. I was told it is not advisable to operate, that there is too much blockage at too many places, and I was told that all they could do is medicate me until I pass on (he used a kinder term).

I was living with angina pain every day, hardly able to go on. In January 1997, I started the Tower Pauling therapy product Heart Technology. After about a week and a half I noticed I did not have any more angina pain. (I have had a little angina pain one time back in April, 1997, I believe the reason for that was because I cut back on the dosage too soon.) I increased it back to the usual dosage of 6,000 mg vitamin C/lysine each per day, and after that I have had no more problems. I have cut back to some amount over 3,000 mg daily after 6 months and don't seem to have any problems. Praise the Lord! I feel like a different person, I forget that I ever had heart problems.

In December 1997 I took the "Nuclear Cardiac Stress Test" and had no sign of chest pain. It seems the Cardiologist lost interest in my case (when the goose that laid the golden eggs died). All he told me after I finally tracked him down was, "Everything is satisfactory, come see me in a year." I have not been back to see him since.

I am convinced that the Pauling therapy as formulated in the Tower Labs "Heart Technology" drink saves lives.

Yours Truly,

Eli Raber

Mr. Raber is no longer with us and we are glad that his last years were without pain. We do know that people often cut back on the vitamin C and lysine they are taking after they feel well for some time.

Health insurance does not cover vitamins. The people who adopt Pauling's therapy pay for their own supplements. A recurring theme in this chapter is that a formerly critically ill patient feels cured, stops taking vitamin C and lysine, and then suffers a relapse after roughly six months.

For nearly eight years we were unaware of anyone who did not respond in a favorable manner, even if he or she had been in a terminal condition. However, during the past several years there have been failures that we blame on the newer medicated stents which cardiologists insert into coronary arteries and bypass grafts. These new stents are causing pain that cannot be eliminated by vitamin C and lysine. We believe that this severe angina pain, which some patients have compared to the feeling of an elephant stepping on the chest, is being caused by the stents themselves (see the case of *SteveFromFlorida* in this chapter).

To our knowledge, there have been no sudden deaths immediately after starting the Pauling Therapy. By comparison, in 1996, before we knew what we know now, one of our relatives died from a stroke during heart surgery.

The following cases are important for various reasons. Several have allowed us to use their actual names because they know how incredible their stories sound. We thank these brave souls who have sacrificed their privacy to help inform the world about the Pauling discovery. To date they remain willing to discuss their "near death" experiences with others, though a few wish to remain anonymous for privacy reasons. At the time of this writing the authors of the following testimonials are still alive, except perhaps the anonymous *SteveFromFlorida* who reported that vitamin C and lysine did not help his condition. There has been no contact from *Steve* for six months.

CAROL SMITH (Longview, Washington, 2001 - present)

Carol Smith's doctor told her more than 12 years ago to make a Will because she was not long for this world. She had great difficulty walking across the room, for doing so generated severe chest pain. When she first sent Owen an e-mail in 2001, there were a few typos and his initial impression turned out to be entirely wrong. When they finally spoke Owen found Carol to be intelligent and very well spoken, with a voice that sounded like a television or radio personality.

The Carol Smith story now spans more than seven years on the Pauling therapy. It is a story of being "totally cured" and then stopping the Pauling therapy, suffering another heart attack, then restarting and feeling "cured" for a second time. Then, following well-intentioned advice to stop the vitamin C, Carol suffered yet another relapse, her third. If it's true that the "third time's a charm," then we hope that Carol really means it this time when she says that she will faithfully stay on her Pauling therapy maintenance.

All three of Carol's apparent cures were during periods when she was taking a Pauling therapy product sold by Tower Laboratories Corporation of Las Vegas, Nevada. All of Tower's products are drink mixes which provide the recommended potency without fillers. The first product, then called *Heart Technology*™, was formulated according to Pauling's advice and contains high amounts of vitamin C, lysine and proline, plus vitamins E and A and several B vitamins. All of the ingredients contained in the *Heart Technology*™ drink mix can be found separately as pills in the vitamin aisle of any drug store.

At one point after stopping Tower's *Heart Technology*™ the first time, Carol's heart became badly damaged as measured by an EKG. When she resumed her use of the Tower products, she upgraded to Tower's *Ascorsine-9®* formula, which at that time included CoQ10, carnitine, taurine, vitamin K, etc. Carol agreed to monitor and share her medical reports. We had reason to believe that high vitamin C, in conjunction with a high quality vitamin E called Unique-E, might return her EKG to normal. As predicted, Carol's EKGs returned to normal, meaning that any damaged heart muscle that was disrupting the electrical signal in the previous EKG was no longer present. Copies of her "before" and "after" EKGs are posted at http://www.practicingmedicinewithoutalicense.com/carolsmith.

We begin Carol's own testimony below after her third relapse. We hadn't heard from Carol for approximately three years, but in 2006 we received an indication that she was having problems and we asked her for more information. The following three e-mails describe her history since the normal EKGs in 2003. This case provides a powerful seven-year lesson about the Pauling therapy and the problems one might expect after stopping the therapy, even after he or she feels wonderful and entirely cured.

Feb 2007

Hi Owen,

Where to start. The 2 heaping tablespoons of Tower Ascorsine-9 that I took yesterday really did help to stop the tightness when the prescription Isosorbide pill & nitro patch didn't seem to be making much of a difference [reducing the pain] until I took the Tower A-9.

I had been warned by my Heart specialist that the isosorbide pills and nitro patches were only a very temporary fix and very shortly they would not make any difference in stopping the angina. He said I needed an angiogram and one to two more stents put in as soon as possible. I would love to prove him wrong. I made it thru last night and this morning so far with no tightness.

Okay, let's see how good my memory is and update you. First, I am still doing better than when I first made contact with you. (I was then 47.) At that time I was hurting so bad that I could barely walk from my bed to our living room without needing to sit down and rest & wait for the angina to let up. I would be totally exhausted & hurting & very weak and yes, my heart specialist had told me to get my affairs in order that I had 6 months to a year to live.

It's been twelve years since he told me that. Luckily I made contact with you and started the Pauling therapy and found out that I didn't need to hurt any longer, the angina was gone. I felt so much better and went out and got a part-time job to help with all of our bills.

The Tower product worked so well that I believed that I was cured and money was really tight so I stopped taking Heart Technology and about 6 months or so later had another Heart Attack. At that time we spoke and I started on the stronger Tower Ascorsine-(A-9) and added 2000 iu Unique E. About 3 months later my EKGs were normal.

Well, sitting here doing nothing the Angina has returned and is getting bad, I need to go take the Tower A-9 and see if it will help as it did yesterday. I am radiating tightness from my chest into my shoulders & down my arms, also up into my neck and into my back. I feel very weak, and aching all over. I need go lay down. I will continue with this email as soon as I feel better.

Carol

The above was the second of three e-mails. The next e-mail is the first message Owen received detailing Carol's third relapse:

First Message

From csmith Fri Feb 23 14:55:40 2007
Date: Fri, 23 Feb 2007 12:55:32 -0800 (PST)
From: Carol Smith
Subject: Re: Heart Technology
To: "Owen R. Fonorow"

Hi Owen,

How am I doing, is a very good question. According to my heart specialist that I saw 5 days ago, not good. He was ready to put me in the hospital and do another angiogram and put another stent or two in. I refused and he put me on ISOSORBIDE 40MG TAB twice a day plus 0.06 nitro patch every morning to be taken off at bedtime. He said it was a temporary fix and gave me two weeks until he said I would be back in touch agreeing to the stent. I believe that the prescription isosorbide is supposed to dilate all my arteries, veins, etc. to get more blood circulating to my heart and stop the chest pains.

According to my heart doctor, the medicated stent he put in was the wrong size, but it was the only size he had to use to save my life. That was last July (7 months ago). He was very proud of how well he did with getting it in place. The problem now is that the ends of the stent are putting too much pressure on the artery and one or both ends are trying to close off and the only option we have is to put another stent in at the ends of the small stent. He said I am running out of options of what they can do. Bypass is not an option.

I am sitting here with my heart racing (pounding), tightness, shoulders and arms aching, nausea, pain in my back, and very shaky waiting for the isosorbide to work. This is basically what I deal with every day. Some days are worse than others; some days I can go to work and other days I am hooked up to oxygen.

I have been out of the Tower Ascorsine-9 (Pauling therapy) product until a friend stopped by with a jar just before I started this email to you. I immediately mixed two heaping tablespoons of A-9 in a glass of water and I am waiting for it to help. This A-9 jar is 9 months old but I am hoping it might still help. I know how old it is because I had given it to her for her husband, he never tried it. I figure it is not as powerful as it would be new but I know it will help.

It has been about an hour now after taking the A-9, and the shakiness, racing, pounding are gone. Still have a little nausea, tightness and aching in my shoulders, but I am so much better.

I gave notice & I am [*personal data deleted* - basically, money is tight]

I thought I remembered reading that having a medicated stent & taking the Tower products would be a waste of money. I had fought against stents & medicated stents every time my doctor mentioned them and always said NO but my family said yes to save my life. I am very glad to be alive but I need help.

I know that Tower Heart Technology works but do you think it can still work as well since I now have a medicated stent in? It definitely seems to be helping today with my heart.

I am going to end and go lay down and rest and hook up the oxygen and see if I can get the rest of the tightness and aching to go away.

Sorry this is so long.

Take care,

Carol

Latest Message

From csmith Sat Feb 24 20:55:53 2007
Date: Sat, 24 Feb 2007 18:55:42 -0800 (PST)
From: Carol Smith
Subject: Re: Heart Technology/Failing Health
To: "Owen R. Fonorow"

Hi again,

The A-9 made a difference and helped stop the really bad angina I had this morning. I did end up with diarrhea from the A-9 but it was worth it to stop the pain. I took the isosorbide pill, Nitro patch, oxygen and the A-9 to stop it.

As far as my history, in January 2003 had 3rd heart attack and we talked and you told me about A-9 & Unique E. After 3 months on these I had an EKG done and it showed Normal.

I continued with the A-9 and the Unique E until the early part of 2005 and life was very good.

I had been in contact with Matthew from A. C. Grace who sells the Unique E product. He was interested in my EKG results. Over time he shared with me ideas about healing your complete body by using special herbs & supplements. I believed what we had done for my heart with the A-9 & Unique E was wonderful and I wanted to have my whole system, liver, kidneys, adrenal, lungs, heart & colon as healthy as possible. So I started following a program to heal that he recommended. I was taking about 50 pills a day which did not include my meds. I wanted to get off all my meds and believed it was possible.

At that point something had to give and I stopped taking the Tower A-9 and Unique E until we finished the program which ended in the summer, 2005.

I felt really good, had more energy than I knew what to do with. But by then I was totally burned out on taking pills & supplements. And my husband said I had to stop spending so much money on all these supplements and since I was feeling GREAT I did not anticipate a problem.

Since I had gotten out of the habit of taking A-9 faithfully, I took it hit and miss. YOU would think I would know better. But I really believed that with everything I did up to now that I was in really good shape and my heart was totally healed.

In March 2006, I started having angina again but this time the A9 didn't seem to make a difference. My Heart specialist suggested an angiogram procedure and possibly a stent. I said NO. So then he suggested the EECP 35-day program along with Cardio Rehab 3 times a week, which was what I was doing until July 2006 when they put a medicated stent in to save my life.

I had not been taking the A-9 since it hadn't seemed to be helping and the EECP staff had asked that I not be taking Vitamin C and supplements during the 35 day program and I was following their & my Doctor's instructions.

I have only one stent in and no bypass surgery. I have fought for the last 12 years to not have any surgery or stents put in. Bypass surgery is not an option for me according to my heart specialist. He said he is running out of options for what he can do for me. I think I have heard that one before.

By the way, I have had lots of dental work done thru the years and 3 or 4 root canals, etc., so I am concerned that they can cause toxicity you warned me about. Last year when I finally had 100% duel coverage I maxed out and had as much as I could done & replaced with the new non toxic products.

I also have on hand Omega 3-6-9 flaxseed oil and I buy organic flaxseed from the health food store and grind it up and use it on all my green salads & food but I ran out of CoQ10. Sounds like I need to buy some. It was so expensive at our health food store.

All these supplements add up and pretty soon especially with the cost of my heart meds, there is just not enough money to cover it all.

Carol

What follows is the original communication between Carol and Owen which began in the year 2001 and has been posted in the Pauling therapy testimonials:

July 23, 2001

Dear Owen Fonorow,

According to the Doctors the main artery is very bad and they cannot do a bypass. Two years ago the main artery closed and I had a 2nd heart attack with damage this time. My heart was in the process of doing a bypass on the blockage when the heart attack happened. The Doctors said that was what saved my life but I would have to learn to live with angina for the rest of my life. After taking the Tower Heart Technology for 3 weeks, I have more energy and no chest pains. LIFE IS GREAT. I ran out and I want that feeling back. I hate chest pains.

Now I just need to get & keep enough Heart Technology on hand. I started with 1 Heart Technology jar and then ordered a second one on autoship. I know that it is helping and I was taking 1 heaping tablespoon every 12 hours. I am almost out and have cut down to a tablespoon a day until my next order comes. (I hope it is soon.) I am dealing with angina again since I cut back on the Heart Technology and would like to go back to two glasses a day or maybe 3. I know I can tolerate 2 heaping Tablespoons a day, haven't tried 3 yet.

Thank you and advice or suggestions are appreciated.

Carol Smith

* * * * * *

July 2001

I woke up this morning with tightness & angina pain in the chest (no fun) This product really does make a difference. Back to the Nitro patches & Nitro pills until I can get on the Heart Technology again. I AM OUT Please send out my next order as soon as possible. Sending 3 jars a month would be fine. Thank you for getting back to me so soon. Yes, you can post this on the testimonials page.

Carol Smith

* * * * * *

August 3, 2001

Owen

I am hoping that the angina pain and tightness will go away now that I am on the drink mix again. I was out of the drink mix for almost 10 days and the angina came back very bad. I had to go back to using the nitro patches daily and some days the nitro pills. I did buy and start taking the lysine, Vitamin C, & E twice a day as pills, but I still had to use the nitro patches. The pills helped but did not totally do away with the chest pains, tightness.

By the way, I called and got the results of the Lp(a) test. I scored a 10 and that was only after being on the Tower Heart Technology for 2 weeks. My Cholesterol is 159 and bad LDL chol is 98. (Doctor is very happy with the results) but they still do not know how to make the angina go totally away. They said I am on as much medicine as they can safely prescribe. I will just have to slow down and live with it. They just don't know about the Pauling therapy and what good it can do. I am going to take 2 level teaspoons 3 times daily.

Thank you for all your concerns and help.

Carol Smith

* * * * * *

Big, big mistake..."I felt so good I stopped taking the therapy."

March 11, 2003

Dear Owen,

I have some really great news to share with you from my appointment with my Heart specialist. He informed my husband and me that my heart is in better shape now than it was 3 1/2 years ago. He told me to keep doing whatever I was doing because it was working. Guess what I had been doing? Using the Tower Heart Technology (Pauling Therapy) product. It REALLY WORKS.

There was such a great improvement that the doctors who reviewed my file with my specialist were amazed. The damaged main artery that had closed during my 2nd heart attack had opened back up and the doctor said it was no longer damaged and is healthy.

Let me explain, I had my first heart attack 8 1/2 years ago. I was very lucky I was in the hospital when I started having a massive heart attack. The left main artery had a 90% blockage and decided to close. When they had stabilized me, they performed a Heart catheterization. It must have been serious enough because the next thing I knew I was in an ambulance being transferred to the best Heart hospital in the area. They performed the balloon process and then had to (rotor-rooter) the calcified blockage. Three weeks later scar tissue formed and the blockage came back and I started having terrible chest pains.

I was brought back in and they ran some more tests and the Doctor informed me that the left artery was so bad that they could not do a bypass. He said we needed a miracle and to start praying, that it was in God's hands and I needed to get my affairs in order and that I had approximately one year or less to live. I got my miracle. My heart started doing a bypass around the blockage by creating collaterals (tiny veins).

Three and 1/2 years later on June 5, 1999, I had my 2nd heart attack, but because my heart was in the process of completing the bypass around the left main artery blockage, the Doctor told me that was what had saved my life. I had all these collaterals providing blood to my artery.

It was at this point in my life that I found the Tower Heart Technology product (based on Linus Pauling's recommendations). It has made all the difference in the world. After taking the Tower Heart Technology for 3 weeks, I no longer had any chest pains and I also had more energy and I was feeling great.

Well don't make the mistake that I made. I was feeling so great and memory of the heart attack and chest pains had vanished. I could do everything I had been able to do before the heart attacks so I stopped using the Heart Technology product. Big Big mistake.

On 1/24/03 I had my 3rd heart attack. I had been doing really great while I was on the Tower Heart formula and I truly believe that if I had continued to take the Heart Technology I never would have had this 3rd heart attack.

Since my last heart attack, I have been in and out of the hospital with chest pains, not able to go back to work, tired all the time, taking lots of medicines and wearing Nitro patches 20 hours a day and taking oxygen when needed for the chest pains. The doctors found that I had 2 blockages and I was put on a waiting list for a Medical Stent. THE DOCTORS SAID THAT WAS MY ONLY OPTION. From my research Medical Stents are not the answer. The Pauling Therapy Heart Technology products are.

Well I am now back on the Tower Heart Technology product it has only taken 3 weeks, the chest pains have stopped, I have energy, I have stopped using the nitro patches and oxygen and I have returned to work 20 hours a week. I am now on a low fat diet and exercising on our treadmill 1 to 2 miles a day and NO CHEST PAINS. This product really works!!!!! and I will never ever stop using it again.

Owen, you and Dr. Linus Pauling's recommendations/Tower Heart Technology product have saved my life.

Thank you

Carol Smith

* * * * * *

EKG Reversal on Unique-E

June 19, 2003

Hi Owen,

Just a note to let you know that I saw my heart specialist and he said the EKG showed no evidence of any heart damage. Amazing isn't it!! You were right. I will email you more later, have to leave for work in a few minutes. Life is great.

Carol Smith

RICHARD

Richard has a long history of heart disease and does not want his full name disclosed for privacy reasons. (He was able to go back to work for a time and is worried that his prior medical history might disqualify him from his normal employment.) He is still alive, but has recently suffered a relapse. He has been diagnosed with congestive heart failure, and only a small percentage of his heart is said to be functioning. To my knowledge, Richard's testimony is based on vitamin and amino acid pills; he did not use the Tower Pauling therapy drink mixes.

March 13, 2006

This is my story about the article that describes what I am taking and doing now to stop chest pain and to stop what would be my fifth and possibly my final heart attack. I have already had five heart operations including a quad bypass and various stent operations, none radioactive thank goodness.

Although I was doing everything expected of me, i.e., following the Ornish plan which in my case involved being a strict vegetarian (vegan) for five years. (I did all this immediately after my first heart attack and quad bypass in 1995.) I was 41 years old in 1995. I continued to get worse.

I suffered three more heart attacks and four more heart operations, the last heart attack (my fourth) and another heart operation (my fifth) in DEC. of 2003. At that time my doctors recommended the possibility of a pacemaker/defibrillator be installed. They did not know what was killing me, but said they would be able to 'take care of me' with more operations! I declined to have an 'operation/test' that would result in maybe having a pacemaker installed. That was Dec. 2003.

Last October of 2005 I did intensive research (on the Internet) to discover what I could do to stay alive. I was having more chest pain and realized I was overdue for another heart attack and operation (based on my previous ten years experience).

I discovered, or rediscovered, what Linus Pauling had to say about heart disease. The 'powers that be' did a good job of influencing me to ignore Linus Pauling, describing him as a failing old man with a 'mental' problem who had once been a great scientist. They said he was wrong about vitamin C. I believed them.

I recently read all of Linus Pauling's books. I was very impressed and I finally understood my situation. This led me to search for more info on Dr. Rath, etc. I now consider Linus Pauling to be the greatest, most significant scientist of this era. He really devoted a big chunk of his life and made the best effort to get the word out about ascorbic acid. Fortunately for mankind he was not alone. Pauling was preceded by great men and women in the study of vitamin C and that work continues with great men and women today.

The real issue now is how to get the word out about ascorbic acid. In my opinion Owen Fonorow may be making the greatest effort for this positive change with his website (along with other folks and the folks that post there). I point friends to his site often.

The following article describes what I am currently doing in terms of vitamin intake, etc. I did slowly stop my statins and blood pressure meds, etc. back around Nov. 2005. I started ten grams of vitamin C every day spaced out five times during the day about that same time. I felt so much better after a few weeks that I continued my research and the use of C and now follow this protocol in full: http://practicingmedicinewithoutalicense.com/protocol.htm

I have not felt this good in over twelve years! I would also like to say that I have read some Atkins. He is one of the few doctors that recommends vitamin C and makes good reference to Linus Pauling. I was getting to be on the heavy side and decided to lose about thirty pounds to get to a weight that makes me feel and look even better. I have already lost thirteen lbs in one month. I don't feel hungry, ever, and I feel better by eating this way. I'm glad I did the vitamin C protocol before I did Atkins. I can say I definitely had my chest pain completely disappear with the vitamins as well as my thoughts of having another heart attack and heart operation. I will continue with vitamin C, etc.

One last word.

The effort to prevent the knowledge about ascorbic acid (vitamin c) getting out to the mass of humankind is in my opinion the greatest tyranny of all time. In closing I wish everyone life, liberty, and the pursuit of happiness,

* * * * * *

Update Nov. 28, 2006 (almost one year after starting the Pauling protocol)

Background: 1995 first of four heart attacks, quadruple bypass, stent Operations, five total operations to date. I declined a pacemaker implant (TEST) in Dec. of 2003.

Continued problem: from 1995 to 2003 numerous heart attacks, operations, feeling almost hopeless. I had tried numerous things that my doctors recommended. I became a strict vegan, no: eggs, chicken, fish, etc. Did not smoke, etc. These practices were put into place immediately after my first heart attack and quadruple bypass in fall of 1995 for a period of over five years.

On the edge of death: chest pain, no energy, feeling of doom, fall of 2005. I did not think I would be alive on Christmas or new year of 2005. My health wasn't good, but I was in a steep decline at about the end of '05.

Rediscovery of vitamin C: I searched the Internet for anything that would help me stay alive. I was desperate, too sick to work, no health insurance.

Course change: Pauling protocol. I don't know if I rediscovered Pauling before or after I discovered the Vitamin C Foundation web site (vitamincfoundation.org). I ordered all of Pauling's books at the local library and read them all. This was about the beginning of 2006, or the end of 2005. Many years ago I remember seeing Pauling on TV and hearing about his work with Vitamin C. I also recall the media in general as labeling him in a very subtle way a "kook". This probably caused me to not read his work back in those years.

I started taking ten grams of vitamin C every day, spread out five times during the day, about the time I rediscovered Pauling and read his books on vitamin C. I continued to find everything I could about vitamin C. I learned on the Vitamin C Foundation about other supplements. I also discovered the rich history of vitamin C research. There were many people I read, including Dr. Rath. I was soon taking as much vitamin C as I could tolerate, 18 grams, plus I added some other supplements. Being a very skeptical person I only started the Pauling protocol with ten grams of C and nothing else. Then I added all the ingredients.

Biggest surprise: I was doing very hard physical work for over two weeks just recently, lots of pain in muscles from head to toe...but NO CHEST, or HEART PAIN!

Recent blood pressure: 122 over 72, a big surprise and I thought it was a mistake. I had just had a big dinner with friends a few hours before.

This may have caused the reading to be lower. My "usual" bp is around 135 over 79 and this is continuing to drop, without any bp meds, or any heart meds in almost a year. I would like to see it 120 over 80. My resting pulse is still high at 100. (My records indicate a bp of over 147 over 95 while previously on bp and heart drugs.)

Weight: I have been of average height and weight all of my life. I am now fifty two years old. Not until I read Atkins and started eating accordingly did I feel really good with no hunger between meals or fatigue, like I always used to experience. The major benefit seems to be a drastic reduction in hydrogenated oil, etc. found in bakery and prepared food.

Other body changes: My skin has cleared up and looks great, I had a small amount of acne on my face for most of my life. My joints are less painful and more flexible. I feel much stronger and better than I have in over twenty years!

Diet: Two eggs for breakfast almost every day with some meat. I minimize carbs due to reading and understanding the Atkins books. Most of my carbs now are in the form of fresh green veg. or fruit. I tried his diet and feel much better. I eat lots of meat, some fresh veggies, and lots of water. I had given up refined sugar many years before my first heart attack, but still had enjoyed sweet carbohydrate deserts.

Bad habits: Although I no longer eat cookies and other bakery, I still love pizza and ice cream. I will overdo it on carbs on occasion and feel worse for it. I read labels and try to avoid hydrogenated oil (check your pizza label) like the poison it is.

Future: Health is great and getting better! I feel physically and mentally great! I think I'm going to be ok!

Thanks everyone,

For health and freedom,
individual freedom,

Richard

JEFF FENLASON

Jeff Fenlason has allowed us to use his real name. Jeff is a veteran and he had been under the care of cardiology for 10 years with all expenses paid. We have mentioned Jeff's case before because (a) he was on his death bed a night or two before he started the Pauling therapy, his children having driven all night to be with him, and (b) the only alternative to orthodox cardiology that Jeff adopted was the Pauling vitamin C and lysine therapy, immediately at the recommended dosage.

July 11, 2001

My name is Jeffrey Fenlason. I'm 55 years old and I have had Cardiovascular/Coronary Artery Disease or CVD for 10 years. Within two days after starting the high vitamin C/lysine Pauling Therapy (14 g vitamin C) I felt like a new man. Here is my story.

I was first diagnosed with CVD in 1990 at the VA hospital in Richmond VA. In 1991, four of my arteries were by-passed. After the operation I was in worse condition than before it; I had angina daily and took oral nitro almost every day. The angina lessened after a year and it was possible for me to go days without the nitro for almost 7 years. In 1998, after moving to North Carolina my chest pains returned on a regular basis so I went back on the pills and then was on a daily time release Nitro. I had resigned myself to the fact that there was nothing that would improve my condition even though I was on several medications to help relieve my angina.

On June 20, 2001 I had three episodes of severe chest pains and went to the hospital. They kept me overnight and ran the usual tests to determine if I was having a heart attack. (I had been through this experience many times in the past.) The next day I went home with all new meds and an appointment for stress test the following Tuesday; I was also placed on a Nitro patch .4 mg an hour that I could only wear for 12 hrs a day. Within an hour of removing the patch I would again need to take oral Nitro pills.

I was back in the hospital early Monday morning, I was in constant pain even after several oral nitro pills. After the first test the doctor told me I was having a heart attack. I was admitted to the ICU and set up for a heart cath the next morning. After the cath the doctor told me that two of my four bypasses were entirely blocked and that the disease had progressed to the point that it was inoperable. Also, the third graft was partially blocked and he wanted to do the balloon angioplasty to try to open it. A deep depression sank in immediately. I stayed in the hospital until Thursday and then returned home with all new meds. Now I was on 50MG TOPROL XL, 60GR ISOSORBIDE ER, 50MG ALTACE and 40mg of LIPITOR. You can look these up online, I did so Friday morning and I wasn't impressed with what the doctor ordered to treat my condition. While sitting there at the computer, feeling really defeated, I ask half out loud, "GOD IS THERE ANOTHER ANSWER! THIS CAN'T BE HOW I AM SUPPOSED TO SPEND THE REST OF MY LIFE."

That is when I did a search for CAV; the first site I went to was www.paulingtherapy.com. I called the 1-800 number after reading everything on the site and left a message. Within a very short time I received a call from Owen. He explained to me the Linus Pauling Vitamin C/L-lysine therapy that is contained in the Pauling lecture. I started taking the cure that day.

I was to be in Charlotte for the balloon on Monday morning, but I overslept (for the first time in years I will add) and was awakened by a call from the hospital. I was already an hour late and they wanted to know if I was coming or would reschedule. I told them I would reschedule. Then I went back to sleep (another thing I hadn't done in years). Again the phone woke me. This time I was told, if I could make it in less than an hour and a half they could still do the angioplasty procedure. I thanked them for the offer, but told them I would rather reschedule. I was then told how important it was for me to get this operation done. I still haven't called to reschedule.

Within two days after this call I felt like a new man. My friends started saying things like, "WOW you look great" or "Man I haven't seen you this active in months." I had previously stayed home a lot when my wife went visiting. The most common words out of my mouth had been, "I'm tired." Well, since taking the Pauling therapy I am not tired all the time. After only a week I stopped taking the Meds that the doctor had prescribed when I left the hospital. It has only been 12 days since I started and I DO NOT HAVE ANY CHEST PAIN. If that isn't proof enough for anyone with CVD to start taking Pauling's therapy, then check back as I will be updating this report often. In the mean time, GOOD LUCK!

P.S. - Another startling occurrence in the last week, I have had joint pain in both hips for years and I have tried many treatments for the pain. It was just difficult walking around Wal-Mart. Well, for the past 5 or 6 days I have had no pain in either hip and I am walking quite a bit. And, no, I am not taking any type of pain pills.

Jeffrey Fenlason

* * * * * *

July 17, 2001

Hi everyone,

This is Jeff Fenlason again, and this has been one fantastic day. I started working on our lawn tractor, it has been out of service for months with a broken front spindle, at 8:30 this morning. I got it fixed and started mowing about 10:00. It felt really good to be out in the yard working, as this is something I haven't been able to do for some time now, hence the broken tractor sitting for months. Well, we have about two acres, one of which is pasture for the horse. We have an electric fence around the pasture, which has been turned off because I couldn't use the push mower to mow the grass from under the wire. In case you are unfamiliar with electric fencing you need to keep the grass mowed or it can start a fire in dry weather. Well anyway as I said this hadn't been done all season and the grass was above the lower wire. Also the cows in the adjoining pasture broke through the wire since it wasn't turned on. After mowing with the tractor for about an hour I decided to use the push mower to finish cleaning out the grass. This was three sides of an acre lot, enough to tire a healthy 55-year-old man, but I mowed most of it myself. Most people wouldn't think twice about mowing their yard let alone post it on the Internet, but remember, I couldn't walk to the barn and back less than three weeks ago. This is really unbelievable for less than a month.

I felt so good that it occurred to me that one or more of my blocked arteries must have cleared considerably.

I thought I would try to shock my wife a little. I told her I had something to show her and asked her to sit on the back steps. She did and I walked across the pasture to where the lawnmowers were, several hundred feet. I then turned around and jogged back to where she was sitting. The look on her face was priceless to say the least. When I got to where she was she said "You jogging?" and then took my pulse, it was 60 beats a min. I was a little out of breath but no pain. Who wouldn't be out of breath jogging across a pasture in NC at noontime in July?

I have a feeling there are going to be a lot of unemployed cardiologists.

About 4:00 we took a shower and drove to the lake to go fishing. I have been several times in the past two and a half weeks. I am really enjoying the walk into our fishing spot. We returned home around 9:00 and had a light meal, including catfish. We spent about an hour sitting outside and talking about how great it was to be able to do things together again. Then we did what any healthy married couple does, sorry, I felt this was important or I wouldn't have mentioned it. As I said, what a day. Remember, I had a heart attack just three weeks ago.

THANK GOD FOR LINUS PAULING!!!

Jeffrey Fenlason

* * * * * *

January 29, 2008

Owen,

You may want to add to my story in your book. Next month I will be 62 and I feel and look better than I did when I was 50. I am more active than I have been in years. People often tell me how much younger I look than my age. Very soon I will be remarried. My fiancée is not yet 20 years-of-age, we are very much in love and we have been going out for 2 years. She wants to have children, and after months of saying no I have decided to become a start-over-dad at age 62 and plan to live to see my children grow.

If a man in the shape that I was in can turn his life around in just a few years, then those who are not on their death bed should be able to achieve even more with vitamin C and a few life changes. Please be advised that we have decided not to become a media attraction.

Jeff

LES (United Kingdom)

Les from the United Kingdom is a member of The Vitamin C Foundation forum. Les's story is another "It's the dosage, stupid" account. Les had been diagnosed with heart problems and after about 18 months of doing well on a good Pauling therapy protocol including 60,000 mg of vitamin C daily, Les decided to purchase the Tower Ascorsine-9® drink mix in order to avoid as many pills as possible.

At that point Les reported that he began to feel worse and was having symptoms of adrenal insufficiency. It took the forum awhile to figure out that after switching to the Ascorsine-9® Les had stopped taking his other vitamin C, without considering the dosage ramifications. Les had dropped his vitamin C intake to 6,000 mg (in two servings of Ascorsine-9®). After increasing his vitamin C to his tolerance level of 60,000 mg daily, Les reported feeling much better.

* * * * * *

June 7, 2007

Re: Update on my heart condition.

Hi guys,

I have now reached the point where I am not progressing any more, I am still on the ascorsine-9, Unique E, CoQ10, cod liver oil capsule, magnesium citrate, multivitamin & minerals, l-arginine, B- complex, hawthorn berries, white willow bark, & 1 beta-blocker twice a day. I also juice everyday and also drink some kombucha, & do a little walking,

On Sunday I ended up in the a&e (emergency room) and thought I was having another heart attack. All the tests were ok, apart from the ecg, which was a little different from the one I had in 2004. The docs reckoned everything was ok, but I knew there was something going on with my ticker, so I asked to have the 24 hour ecg monitor. I have an appointment to be fitted with one on the 25th of this month. If everything turns out to be okay, the problem will most probably be stress.

But, I still feel a lot better than when I was taking the drugs which the hospital prescribed for me. The docs tried talking me into taking an 80mg statin even though my cholesterol is not high. I have always refused them & did so again this time. I told the docs I was taking nature's statin as recommended by Linus Pauling. I tried to explain his therapy to them but they were not interested. Out of the 4 docs I saw, only one had heard of Pauling. That says it all.

Les

We asked Les for an update on his symptoms to try to find out what had put him in the hospital, and below is his response:

Update:

My symptoms are palpitations, adrenaline release, sweating, getting very tired very, very easily & occasional mild pain & also still tired after sleep.

No mercury or root canals, I have read Dr. Levy's book which another forum member kindly sent to me.

I am always open to any suggestions. I am not a medical person & all my knowledge about alternative medicine comes from this site, Dr. Mercola & other docs on the web plus I read whatever books I can get hold of. I'm on my own here as there are no naturopathic docs in the UK as far as I know. All the docs I see have closed minds. My gp will not even have a discussion about vitamins, most probably knows nothing about them.

Les

We then asked Les for a reminder about the nutrients he was currently taking and he responded as follows:

Update:

My daily dosage of CoQ10 is 150mg, I will increase it to 180mg a day.

I will visit my gp to obtain my blood sugar reading.

I have found a good omega 3 oil. It is double distilled & then concentrated & enriched with extra omega 3. Each gelatin free capsule provides130mg of omega 3 DHA as well as 200mg of EPA.

My daily tolerance for vitamin C has always been 60 grams in 4 x 15 gram doses, have started taking extra 500 mg ascorbic acid pill every 4 hours.

I have started my daily record of how I feel.

I really appreciate the inputs from members of this forum. Without this forum I would most probably still be in the dark, and still taking the big pharma drugs.

Once again I would like to thank you and all the guys and gals on this forum for the good advice and help that I have received over the last 18 months or so.

Les

* * * * * *

Update Reply:

Thanks. I have just read Seymour's post, will have a try with the chondroitin sulfate, as I do have some scarring from 2 heart attacks. On the 25th of this month I'm having a 24 hour ecg monitor and I will make sure that I get a copy of the tracing for future reference. Since changing my gp some of my medical records get lost/misplaced, so by getting a copy I will be able to get future tracings checked against my old ones. Hope springs eternal. Thanks again,

Les

* * * * * *

Update

Owen,

At the moment I am taking 1 heaping teaspoon of

Tower Ascorsine-9 twice a day,
5 unique vitamin E in the morning,
1 multivitamin & mineral,
1x100 mg magnesium citrate,
1x500 mg l-arginine
180 mg CoQ10

I was taking the 60,000 mg vit c before I started on the Ascorsine-9.
I increased my CoQ10 from 150 to 180 mg yesterday, tomorrow I will increase it to 300 mg.
At the moment all my vitamin D is coming from 20/30 minutes in the sun.
I am not taking any vitamin A, first thing tomorrow I will nip out & get some vitamin A & vitamin D3.
I was taking 1x3.125 carvedilol twice a day, I have reduced it to 1x1/2 3.125 twice daily. I will not bother to consult my gp over reducing the carvedilol, we do not see eye to eye. I fell out with him over the treatment my wife received, I only go there when it is really necessary.

Les

At this point someone at the forum noticed that Les had been taking 60 grams (60,000 mg) of vitamin C, but that he was now only taking two servings of the Tower product, or 3 grams of vitamin C per serving! Obviously, if his vitamin C intake was down to only 6 grams and his tolerance was 60 grams, then his adrenal insufficiency symptoms were explained by too little vitamin C. So we wanted to verify exactly how much vitamin C Les was taking.

Update:

Yes, I stopped taking the 60 g when I started on the Ascorsine-9, & now the only extra C I take I started just recently. I have started taking a 1x 500 mg tablet every 4 hours. So should I go back on the vitamin C powder again? Taking the A-9 I was under the misconception that and the 500 mg tablets were enough, so now does the forum think I should go back to bowel tolerance?

Les

We told Les that we thought his drastic drop in vitamin C is what had caused his problem. He had lowered his vitamin C intake from 60,000 mg to 6,000 mg daily! We suggested that because of his unusually high bowel tolerance, he should add 1 to 3 teaspoons of vitamin C powder to the Tower A-9 drinks, as this adds 4,000 to 12,000 mg vitamin C to the 3,000 mg in a serving of A-9.

Post update:

I agree that this forum is a lifesaver. I will follow its advice to the letter. I do get some gas now & again. No problem, a small price to pay. Thanks all, really appreciate all your input.

Les

* * * * * *

June 13, 2007

Owen,

I am now taking 15 gm vitamin C with the Tower Ascorsine-9 twice a day and large amounts of vitamin D twice a day, have increased my CoQ10, have lowered the beta blocker dosage by 50% and am feeling 100% better!! I have ordered vitamin D3, it's hard to come by in the UK, have also added vitamin A. My C bowel tolerance is still 15 grams per dosage.

It is great to feel well again, a million thanks.

Les

GERALD (Australia)

After an article about the Pauling therapy was published in the Australian Nexus magazine, we began to get many reports similar to the following from Australia:

Hello Owen,

You may be interested in the following information.

I have now been taking Tower Laboratories Heart Technology [Vitamin C/Lysine] for a period of two years. The amount taken has been 3000mg ea. in total each day, split morning & evening. I may have missed on a couple of occasions. These quantities are listed as a preventive dose. Prior to taking Linus Pauling's therapy my condition was as follows.

I was admitted to Hospital with a stroke 24/004/03.

The only damage I suffered was loss of eyesight in the left eye.

Following my discharge from hospital I was subjected to a Carotid angiogram which proved unsuccessful due to the plaque within the arteries. The left artery was down to a trickle with atherosclerosis spreading deeper into my cranium. My right artery was also infected with atherosclerosis.

At this stage I decided on a second opinion. The results were the same. I was virtually told that it was only a matter of time.

At this point of the saga I decided I would navigate my own destiny. "Enter [The Pauling Vitamin c/lysine therapy] Heart Technology"

The only one request to my vascular surgeon would they monitor by scan the effect. This they agreed.

I have been scanned every six months.

Eighteen months has passed and no comment has been made by the Vascular surgeon.

In June 22/6/05 I had my last scan. The result is that the atherosclerosis in the left carotid artery, according to the vascular surgeon, has settled down. The atherosclerosis beyond the blockage, which was entering deeper into my cranium has disappeared, in other words, the artery is clear. My right artery is also clear.

The Vascular surgeon is staggered.

Maybe had I gone onto the therapeutic quantities, my left artery may have cleared the plaque blocking it?

However this is very good news. My next scan is in December so I am hoping for just as good news. Will keep you posted on the next scan.

Regards,

Gerald G.

OTHER REPORTS

May, 2004

My mother was diagnosed with an 80% blockage 18 months ago, but she would only take 1 jar of the Pauling therapy per month. Her last visit revealed that her blockages are all gone. The surprised doctor asked what she was doing - she told him about Tower Heart Technology. His response, "I doubt that could have any effect. On the other hand, I don't have 2 Nobel prizes...."

Anonymous

* * * * * *

July, 2007

On Feb 5th 2007, my Dr. stated the Left Anterior Descending artery or LAD was 97% blocked and the Right Coronary was blocked 67%. A medicated stent was implanted in the LAD. On June 26th the Dr. stated the Right Coronary Artery was completely clear and a 10% blockage in the LAD where the stent was implanted. Tower lab's Ascorsine-9 receives the credit for clearing the Right Coronary Artery. Thanks for a great product that works. The tests used in determining the % of blockage were Cardiac Catheterization, Coronary and Heart x-rays with a dye injected."

J.P.C.

* * * * * *

September 12, 2002

I Just wanted to say thank you. My husband was diagnosed with cardio-vascular heart disease two years ago. He quit smoking (used to smoke three packs a day), we changed our dietary habits and he got very serious with his exercise program. He had a 30% blockage in one artery and a 70% blockage.

His cardiologist tried him on countless medications which seemed to do nothing more than cost us a fortune and cause him to have just every side effect imaginable. All the while he continued to have chest pain.

I cannot tell you how many times in the past two years we have been to the emergency room thinking he was having a heart attack.

In June of this year he was back in the hospital. They put in a stent. The 30% blockage now was 95%. Through the summer he didn't make any improvement. He still had chest pains every day and would get nauseated and lightheaded in extreme heat and/or with exertion or sometimes without extreme heat or exertion. THANK GOD I FOUND YOUR WEBSITE!!!

In August of this year he was back in the hospital. The stent was not open all the way and had scar tissue surrounding it. They cleaned out his arteries and put in another stent. He began taking Heart Technology (and quit taking the medication his cardiologist had prescribed) just about a week or so before the last stent was put in. Yesterday, it was 100 degrees here and he called me from work to say he felt like a million bucks! He had been feeling better the past couple of weeks but, I don't think either one of us were getting our hopes up thinking it was permanent.

I have spent the summer wondering "Is this the day he has the heart attack?"; "is this the time I take him to the hospital and he doesn't come home with me?" God only knows what has gone through his mind!

I believe you have saved his life and I thank you on behalf of myself and our three sons (ages 5, 11 and 17) for keeping him with us for what I believe will be a very long time.

Sincerely,

Mrs. P. K.
Alexandria, KY

Failures

During the early years, results were almost universally positive, much along the same lines as the preceding cases and testimonials. Then we began to notice that not everyone was reporting the immediate reduction in angina pain, and some were even giving up on the therapy. The most obvious problem, aside from inadequate dosage or stopping the therapy, was with the so-called medicated stents.

My problem started the days after the stent was implanted. Just taking vitamin C, I have read is not able to overcome this DES problem. And they are right. So I went in search of hundreds of things, supplements, therapies, surgeries, all over the world, anything to even lessen the pain. Imagine a vice squeezing your chest day and nite, non stop, or like an elephant on your chest, the pain so bad it makes you want to throw up all day long, all starting the first then second (even worse) morning after the implant of the stent, with a bad fever for a month after implantation.

SteveFromFlorida

SteveFromFlorida is an anonymous poster to the Vitamin C Foundation forum. Steve reports being fit until he had a heart attack while playing tennis. At that point a drug-eluting stent was placed in one of his coronary arteries. Steve claims that he has had persistent, severe angina pain that began after the implantation of that stent, and that vitamin C, lysine, and proline have had no effect on reducing this pain. Steve reports that he has tried virtually everything. He even visited Thailand for stem cell replacement therapy, to no avail. His only relief has come not immediately but over time from intravenous phosphatidylcholine and large amounts of oral chondroitin, along with very high amounts of "dry" vitamin E. It is noted that Steve has "gas" from taking vitamin C.

This post is long and emotional but valuable. We have not heard from Steve for six months at the time of this writing.

PC and Chondroitin Sept 2007

Owen,

Currently I do not take vitamin C; it is not helpful for the problem of a stent. It was not helpful for me, though I tried vitamin C and all of the Pauling theory supplements in your protocol, in large quantities. In fact, I felt worse when I did the Pauling supplements that might be okay for one without a stent. I now take phosphatidylcholine (PC) and chondroitin (CH), and a little vitamin E and few other supps. I juice a lot and get vitamins that way, fresh and live. But I take a large quantity of PC and CH for my particular problem of "stent itis"! Namely nothing but pain since they installed a drug eluding coronary stent. Stents are garbage.

SteveFromFlorida

∗ ∗ ∗ ∗ ∗ ∗

Previously…

July 2007

After I received a giant stent sandwich from the Cleveland Clinic 19 months ago, my severe angina started, and I had it every day since.

Stents, especially drug eluting (sirolimus vessel/heart cell killing) cypher stents are terribly defective products, in my opinion, creating inflammation day and night, as they resist (and perhaps kill endothelial cells in) the coronary artery and its natural need to change size and flex. Linus Pauling's theory will not work very well in people with these stents and angina, as pointed out on various web sites. But vitamin E does. I still take some 1 or 2 gm of C each day, but I do now take 3,600 IU of natural vitamin E with food. I make sure my brand is natural and has at least some gamma E in it. Lately I buy dry E, because I also follow Dean Ornish MD and Caldwell Esselstyn MD (Cleveland Clinic heart diet guru) diets and avoid oils when I can. I tried everything before and nothing made a dent in my severe angina until large amounts of vitamin E.

Vitamin E has been the only thing that worked to lessen my angina which was very severe and constant. Now, I take no anti platelet, no blood thinners, just an occasional aspirin 81mg. Sometimes I don't even take that. I stopped Plavix after 13 months. Herbs, hawthorn, terminal arjuna, cayenne, and about 12 other herbs were useless. Beta blockers, and the new drug Renexa, all useless except sometimes Toprol, which is not the greatest drug, but slightly helpful for the pain. I take 50 mg Toprol now. (Before E, I took 200mg Toprol for the pain).

Ace inhibitors did not remove any angina at all, nor calcium channel blockers, nor 35 EECP treatments, all useless to me. I tried 80 mg Lipitor, Zocor, and on and on, useless for the angina. Nothing but sickness and deathly pain. I also tried arginine, gobs of proline and lysine per Pauling, prophytl L Carnitine, Fish oil, Flax and oil, all useless. I tried serrapeptase, nattokinase, and on and on.

I tried everything you see on the net including everything "Life-extension" sells, only getting the same bad angina. I never found success with anything but vitamin E or any product of the hundreds I tried.

Also useless to me was Chelation, and Phosphatidylcholine (PTC) IV therapy, though I think PTC may have been responsible for a recent zero calcium score, but not relieving angina.

Low power laser on the heart was also useless. I even went to stem cell surgery in Thailand (Theravitae/Vescell) which was a complete waste of money (current price $32,000.00 all for nothing).

I had the same exact angina for months when I came back. Never improved. Stem cell therapy did not do anything to my angina whatsoever, though sadly I thought it would. Even exercise and diet were not useful, though I still stick to them religiously (I and a strict vegetarian now).

When I don't take my E for even a day or two, boy do I feel it. I feel sick in the heart, squeezing, inflammation, burning in the heart, very bad.

When I take the E, it cuts it all in half, and I have a much better day and night, though I still have some angina, just less. Probably easing the severe stent/vessel inflammation.

I take a lot of E due to the recent study that indicated that much of it does not get in the cells anyway, and is just excreted. The recent recommendations I read is to take E while actually having your full meal, not alone with say just a drink. Very important. Or else it is just excreted.

Small dosages of vitamin E under 1200 IU do little for my angina.

Many older people who followed the Shute's theories of the 1950's and 60's, tell the same story I am telling. But it seems to get lost in the noise of the recent foolish studies and bad headlines bashing vitamin E. Really too bad.

Pauling talks very favorably of E and its heart benefits (anti-anginal, etc.) in his books. Funny thing though. I tried "Unique E" brand for a month last year, and did not have any of the same benefit as I got with dry E. But "Unique E" is very oily. Perhaps better to avoid the oil like Ornish and Esselstyn say? Not sure why some say it is better to take E with oil.

I found it better to take tons of dry; such an E is almost a food itself it seems to me. In fact I chew the E pill contents with the food in my mouth to make the body accept it as a food, to avoid the problems outlined in the recent article about absorption problems of E. I sure don't believe that recent study suggesting E is useless. Probably because of the silly low dose used (as low as 200 IU) perhaps the low quality brand used (synthetic and Alpha only) or because the pill was just popped in the mouth by the study participants with no meal.

I am a walking lab, and vitamin E sure worked on me. It sure is not killing me. Post if you can about this as I am most curious if anyone has had the same amazing result I had with it. All true no joke.

SteveFromFlorida

* * * * * *

September 2007

I juice 2 to 3 whole cabbages and 3 cucumbers per day, and lots of other things for vitamin C and all other vitamins...and I have 3 glasses of a product from Greens for Life (powder with Pomegranate, gogi, acai, blueberry, and on and on) using giant scoops. I drink this all day long, with as much live juice as possible. Probably getting more bioavailable C than taking ascorbate, which just gave me gas. But just taking vitamin C of all kinds, and all of the Pauling stuff for 2 years, was a waste of time. My angina just got worse. Really. Nothing helped till using chondroitin (8,000 mg per day bovine) and phosphatidylcholine (PC).

Things really started happening favorably when I went for 20 IV treatments of Phosphatidylcholine at my doctor, then followed up with tons of chondroitin (about 8,000 mg per day) for about 6 weeks. Then I started to get relief, though not perfect.

Note that I previously consumed mass quantities of Lysine, proline, ascorbic acid, l-citrulline, and on and on, the year before, all the Pauling stuff, but all a waste of money and time for me... sorry just the facts.

But the biggest waste of time, pain, etc. was to go to the Cleveland Clinic and have them put a piece of junk stent in me, drug eluting... nothing but suffering since. Fever for 31 days after implantation, an elephant type pain in my chest every day for over a year. Near death pain, every day for months, horrible... And they had no answers, just did not want to deal with it. Basically told me to die, consider a bypass, but they did not even know what to bypass, were not sure.

In my opinion, drug eluting stents should be outlawed. Drug eluting stents are something dreamed up by some strange scientist/doctor just trying to make a buck. stay away from them, very dangerous...

I got more benefit from a few IV treatments of PC than anything. Costs about $135 per treatment at an alternative doctor's office. But I also followed up with tons of Chondroitin, not sure if it would have worked well without Chondroitin.

I also take some Gamma vitamin E to avoid oxidation of the PC, but just relying on the Unique E therapy was also useless. Did not work. Though I think important to take some E if doing PC. Most IV PC drips do contain some vitamin E in the mix...

SteveFromFlorida

* * * * * *

Chest Pain, Heart, Phosphatidylcholine, Chondroitin

The reason for all of this is due to the Drug eluting stent. It makes any simple molecular therapy useless. This is even a caveat on the better vitamin C heart therapy products. So what's a guy supposed to do? In my opinion, DES stents are dangerous medical garbage. The bare metal stent was bad enough for the public, but they had to dream up something that releases a drug (sirolimus) that kills all cells in its path downstream from the stent in the subject major vessel, destroying all the microvascular pathway to heart structures, and also destroying parts of the heart structures to boot.

Last week's latest study out of Switzerland proved this occurs. Yet US cardiologists still put DES stents in people, and are expecting to put them in millions of Americans this year. Google the news on the collateral problems of DES stents. I am certain the Cypher and other drug eluding stents are causing problems like mine. Go to PTCA.com, a site sponsored by the stent industry.

There you will find thousands of posts of people with similar problems like mine.

Originally the editor of that PTCA.com site would not post the DES stent complaints. Now with so many complaining, he has no choice. He still will not accept most of my "warning" posts, as his funding is from the stent industry, so I go to other sites like this one trying to warn others of the dangers of the drug eluting stents.

My problem started the days after the stent was implanted. Imagine a vice squeezing your chest day and nite, nonstop, or like an elephant on your chest, the pain so bad it makes you want to throw up all day long, all starting the first then second (even worse) morning after the implant of the stent, with a bad fever for a month after implantation. Vitamin C, proline, citrulline, every amino did nothing to even slightly lessen the pain after many months, useless. Sorry just the facts. And many other people I have read have the same problem. I suppose these things like vitamin C might be helpful if one has simple atherosclerosis, and not severe "stent osis". Really, atherosclerosis and heart disease are very easy to deal with, with enough patience, discipline. But now we are dealing with millions of people worldwide who have metal junk placed in their arteries.

As for Zocor, I am absolutely certain that it so far has not made me feel any better or worse. I have been both on and off statins for various time periods so many times. But I am reading some studies that suggest some anti inflammatory or other benefits, that may be relevant. See also Caldwell Esselstyn's web site, quite impressive (Google his name, very interesting heart stuff). He is the well known vegetarian Cleveland Clinic doctor that got fed up with modern cardiology, and claims to have a cure for heart disease, with angiographic proof. Great site, but so far his theories have not worked on this DES stent victim, and I think his theories are nice if one has no stent to deal with. I spoke to him and his recommendation was to get the LDL way down, even if you must temporarily with Zocor.

So far Zocor is a big no go for my symptoms, though it does lower my LDL (of course, hello). So great guy Dr. Essenstyn, but I don't think his theories are useful to DES stent victims. Thus why statins are my current temporary sub-experiment. Believe me, I know how to go off them. I have done it many times. You chuck them in your favorite toilet, and heavy up on a multitude of other LDL reducing things; there are so many, like PC for example. But not just yet, I want to give Zocor a few more weeks. Really, I would take the devil's medicine if that could help. I take the new ubiquinol Life Extension product to counter the Co Q issues. Seems like a nice product so far, though I don't notice it helping the pain or any other benefit at all. Hope it is not just hype of the same old Co q with a different twist.

I also am experimenting with hyaluronic acid currently to help with inflammation of the vessel, repair of scarring if any on the heart (if any). Also hyaluronic for prevention of cancer. DES stent victims have a ten fold chance of getting cancer of some kind, per more gruesome studies from Switzerland in August 06'.

Not sure if hyaluronic acid is helping me. The only thing that I was doing that ever made visible progress as a DES stent victim, that I can point my finger at, is Phosphatidalcholine and Chondroitin. And I believe with those with simple atherosclerosis, PC and CH will help even more.

Try dissolving metal and polymers, not too likely with vitamin C; and it is impossible to remove a stent surgically. You have to just live with it or die with it. And surprisingly little interest in what problems cardiologists and the manufacturers are causing people. Really, many things can eliminate calcium from arteries, heal atheromas, remove cholesterol and plaque and on and on. That is simple. But unfortunately, there is no chapter in the nice little vitamin C book you are emphasizing that addresses, on how to deal with a great big DES metal stent; a stent that had released poison and destroyed ones microvasculature, and which unnatural rigidity constantly causes painful and dangerous inflammation and thrombosis.

PC seems to be the only partial help. Note: that PC is not a cure, and is only about 40% helpful in reducing the DES stent pain, and must be repeated over and over for one's entire life. So for me it is not a perfect remedy. For one without a DES stent however, IV PC just might be the perfect remedy, especially when combined with CH.

SteveFromFlorida

Why the Pauling Therapy May Not Work in Some Cases

Over the years we have received an increasing number of reports from people who felt that the therapy did not benefit them. Medical science has been negligent, and we can only guess what is happening. Based on our knowledge of the Pauling/Rath theories and the callers' testimonies, we generally attribute these failures, e.g. the lack of reduction in chest pain, to one of the following causes:

1. Insufficient Daily Dosages

Dosages must be adequate and maintained for symptom relief. Ascorbic acid and sodium ascorbate are the only forms recommended. (For example, a calcium ascorbate form would provide too much calcium at the recommended Pauling therapy dosages. We have received failure reports from persons using high dosages of the calcium ascorbate form of vitamin C.)

The following summarizes Linus Pauling's recommended dosages of vitamin C and lysine:

- **Vitamin C (ascorbic acid) 6,000 to 18,000 mg daily.** Pauling recommended this amount of ascorbic acid daily. Generally, the higher the dosage of vitamin C one takes, the more rapid the response. Normal bowel tolerance (see www.orthomed.com/ titrate.htm) ranges from 4 to 20 grams daily. Note: Experience shows that as much as 8,000 mg may be required for six months to lower total cholesterol. Per Pauling's 1986 book, *How To Live Longer and Feel Better*, it is acceptable to add baking soda to ascorbic acid to maintain a balanced urine pH.
- **Lysine 5,000 mg daily.** On video Linus Pauling recommends supplementing 5 g to 6 g (5,000 to 6,000 mg) of lysine for advanced disease. Owen's uncle was a 70-year-old man whose 50-percent carotid artery blockage was reversed with as little as 2,500 mg of lysine (and vitamin C) daily for one month. There is little reason to believe that there is additional heart benefit from supplementing more than 6,000 mg of lysine daily and in fact, 2,000 to 3,000 mg may be effective for many people.

2. Blockages Greater Than 90 Percent

There must be blood flow for the binding inhibitors to reach the plaques. Also, thicker plaques are harder to disintegrate. Great success is reported when blood flow is 50 percent or greater; less success is seen when blood flow is estimated at 10 percent of normal (90-percent blockage); and there is almost no success when the artery is 100 percent blocked.

3. Interference From Prescription Medications

Cholesterol-lowering drugs, called statins, may lead to a form of heart failure called cardiomyopathy.

Many heart medications make patients worse. Specifically, the statin cholesterol-lowering drugs (e.g., Lipitor® and Zocor®) interfere with the body's own synthesis of coenzyme Q10, leading to muscle degradation and heart failure without CoQ10 supplementation.

According to CoQ10 experts, higher dosages of the statin cholesterol-lowering drugs (80 mg) are increasing the cases of heart failure and cardiomyopathy. We believe that muscle destruction causes resistant angina pain in many heart patients — pain that is not eliminated by the Pauling vitamin C/lysine therapy.

Common blood thinners such as Coumadin (warfarin sodium) and heparin block the action of vitamin K to regulate calcium, and this causes rapid calcification of soft arterial tissues.

4. *"Medical" and Radiation-Coated Stents*

Many patients on the Pauling therapy also had bare metal stents (hollow metal implants) inserted into their coronary arteries before they began taking vitamin C and lysine. For many years there were no reported problems. The recipient of almost every bare metal stent who then adopted the Pauling protocol reported "miraculous" results — until recently.

In the past few years there has been a change which may be the result of increased use of the newer, so-called "medicated" stents that are coated with a chemotherapeutic agent that deadens or kills the intima, or inside of the artery.

This approach, as well as the newer radioactive stents (stents with radiation pellets), may nullify the effectiveness of the vitamin C and lysine therapy. The Pauling therapy depends on the artery healing itself — properly. The new stents and procedures interfere with healing. See Chapter 11 for a more detailed description of the radioactive stent issue.

While the newer stents may cause problems and according to most cardiologists cannot be safely removed, the Pauling therapy protocol would still be expected to have benefits throughout the vascular system.

5. *High Blood Sugar/Type II Diabetes*

After a decade of monitoring success with the Pauling therapy we began to receive reports of relapses in Type II diabetics. After initial improvement, these diabetics for some reason relapsed and required angioplasty/stent placement and/or coronary artery bypass grafting. (In one case, intravenous EDTA chelation therapy "worked wonders" for the renewed angina chest pain.)

Combining ideas from various sources including Drs. John Ely and Sherry Lewin, we believe that when blood sugar is high, the blood sugar molecules compete more fiercely with vitamin C molecules for insulin-mediated transport into cells. The cell membrane disturbance would cause trouble for both kinds of molecules trying to pass through the cell membrane in Type II diabetics. Furthermore, ascorbate is broken down faster and less of it becomes biologically available when carbohydrates are present with it during digestion.

The problem of glucose-ascorbate antagonism (GAA) can be overcome by eliminating sugars and simple carbohydrates from the diet. If this isn't possible, carbohydrate consumption should be reduced at the time vitamin C supplements are taken.

According to several alternative doctors, the nutrient R-alpha lipoic acid (ALA) may help diabetics lower blood sugar levels, and taking 300 to 600 mg of ALA may be an important adjunct for people with both heart disease and Type II diabetes.

6. Coronary Artery Bypass Graft Surgeries

The standard heart surgery, coronary artery bypass grafting (CABG), generally bypasses occluded (blocked) coronary arteries using large veins harvested from the legs. These veins are comprised of thinner tissues and are not as strong as the native coronary arteries that they bypass. It is possible that one reason the weaker tissues have worked in bypass operations is because of the tissue-hardening effect of both Lp(a) and the blood-thinning drugs.

Elevated Lp(a) may be of significant value for bypass surgery recipients. The leg veins used for the bypasses are generally weaker than the coronary arteries that they bypass. If the Lp(a) binding inhibitors nullify the ability of Lp(a) to create "nature's plaster casts" after surgery, the bypass might weaken or even collapse. However, too much Lp(a) without binding inhibitors, in theory, will also lead to rapid reocclusion, called restenosis.

Fortunately, those who follow Pauling's advice are less likely to someday require coronary artery bypass grafting, making this a moot issue. Until we know more, in our opinion bypass patients should monitor their serum Lp(a) levels and adjust their binding inhibitor dosages accordingly. The optimal levels are likely to be above zero but not much above the normal range.

There is no reason for bypass patients to limit vitamin C intake. Bowel tolerance vitamin C is always indicated and should not hasten the collapse of a bypass graft. As a precaution, we recommend that only lysine and vitamin C be used — no proline — in early bypass patients where a vein was used for the bypass. Surgeons have recently begun to use arteries from the chest (internal thoracic or mammary artery) or elsewhere in lieu of leg veins for bypass grafts. Because arterial tissue is stronger, normal binding inhibitor dosages, including proline, are probably appropriate in these cases.

7. *Smoking Away Vitamin C - Cigarette Smoking and Dental Toxicity Deplete Vitamin C Levels*

Cigarette smoking and dental work create toxic loads that deplete vitamin C levels in the blood. Cigarette smoking depletes vitamin C levels so much that the National Research Council, in 1989, revised the smoker's recommended daily allowance (RDA) for vitamin C, raising it from 60 milligrams

— the RDA for the general population — to 100 mg. However, research since 1989 suggests that the revised RDA for this vitamin still falls far short for smokers.

Cardiologist Thomas Levy, M.D., J.D., and dentist Hal Huggins, D.D.S., believe that all root canals eventually become toxic and drain vitamin C levels in the body. Levy and Huggins have written a book about dental toxicity entitled, *Uninformed Consent: The Hidden Dangers in Dental Care.* Levy stresses that unless the toxicity is removed, vitamin C supplements will likely be useless. In the following discussion Dr. Levy compares it to fighting a 'stream' of toxicity with a vitamin 'blow dryer':

> First and foremost, and I regret this perhaps was not emphasized enough in the book Stop America's #1 Killer, heart patients must address all chronic dental toxicity, especially root canals and advanced periodontal disease. A blow dryer on a stream of (toxic) water has no effect. You stop the stream, and the dryer can finally take effect.
>
> On the question of root canal toxicity, in our book we mention that of the more than 5,000 consecutive root canals extracted and tested in a laboratory, 100% were highly toxic. The root canal procedure is fatally flawed and assures that the patient leaves the office with an infected tooth that only worsens over time.
>
> I have seen a single root canal cause fulminantly progressive coronary disease, in spite of perfect supplementation. Post-extraction, it stopped progressing and even began reversing almost immediately. — Thomas Levy in private correspondence

Final Note

The Pauling therapy of high-dose vitamin C and lysine is perfectly harmless, non-toxic, and inexpensive. If Pauling and

Rath are at some point proven wrong, there will have been no harm done to anyone by ingesting these substances — substances that in smaller amounts are essential for life.

The case reports in this chapter are but a handful of the hundreds of reports we have received. Many more testimonials are posted at www.practicingmedicinewithoutalicense.com. We'd like to see studies performed using the right dosages, though we are not optimistic that a true double-blind, randomized prospective study could ethically be conducted. It would be inhumane to deprive any human being in a placebo group of the benefits of vitamin C and lysine for atherosclerosis, not to mention that when this theory becomes common knowledge among terminally ill heart patients, they will all be taking vitamin C and lysine on their own.

Even in the cases of angina caused by the new medicated stents, it is not harmful to try the Pauling therapy. Even if the Pauling protocol does not eliminate the pain, it will still improve the health of the rest of the body.

Every person in America — every person in the world — should be made aware of this discovery. We cannot imagine a valid reason for performing another coronary artery bypass graft surgery, angioplasty, or stent placement as long as there is a sufficient supply of vitamin C and without first trying Pauling's therapy.

Chapter 9

Published Clinical Studies Run By Medicine To Test The Pauling Therapy

NONE.

The best evidence that medicine only pretends to be based on science is the fact that medical 'science' has failed to acknowledge, investigate, or publish a single study on Linus Pauling's theory. They have had 20 years in which to do so but have failed utterly, to the great detriment of public health. We believe that the vitamin C theory has been kept under wraps by deliberately failing to run and publish studies that would have been conducted in any other field of science.

Economic factors can distort the science of medicine. We believe that the failure to publish studies is inexcusable. Such studies, long overdue, would have been routinely conducted in any other field of science. In our opinion, studies would have made Linus Pauling a National hero. Instead, his work lies dormant because there is no study to "defend" it. We must keep in mind the caliber of the scientist who made this claim and the utter lack of toxicity, and test this theory no matter how outrageous the idea may sound.

Sadly, this failure of medicine to test the Pauling theory is also harming the great apes in zoos worldwide. The gorilla suffers the same GLO defect as humans, and zoo gorillas everywhere are dying from heart disease.

The purpose of science is to seek truth. Apparently medicine does not wish to follow truth in these matters. Linus Pauling was one of the world's great scientists and, in fact, the world's only recipient of two unshared Nobel prizes. It is unconscionable that his theory and therapy have not been thoroughly investigated. This appalling failure, and the failure at the multi-billion dollar-funded National Institutes of Health, will be explored in great detail in The Pauling Therapy Handbook, Volume II.

One reason for the complete lack of clinical studies is that the so-called upper tolerable limit for vitamin C has been established as 2,000 mg. This limit effectively undermines any effort to study higher dosages of vitamin C. This limit and the underlying RDA are flawed and should be reconsidered in light of the Pauling/Rath theory.

According to professor of pharmacology and vitamin C expert Dr. Steve Hickey, "The US Institutes of Medicine and NIH have not been willing to defend the RDA or the tolerable upper limit for vitamin C. The upper limit is not based on evidence."

We know of an unpublished study with the proper dosages. In March of 2003 we received a report of a three-year study in 200 men in the United Kingdom with the proper dosages of vitamin C and lysine. The study ran from 1997 to 2000. The preliminary report we received from the study author, K. Kenton, explained that atherosclerosis had been nearly halted in the vitamin C/lysine group taking 6,000 mg of vitamin C and 6,000 mg lysine, while atherosclerosis remained severe in controls. This study was completed at least eight years ago but has still not been published.

Dr. Rath conducted another confirming, year-long study with lower dosages than Pauling had recommended.

"At Framingham, we found that the people who ate the most saturated fat, the most cholesterol and the most calories weighed the least, were more physically active and had the lowest serum cholesterol levels." — William Castelli, M.D., Director of the Framingham Study, The Archives of Internal Medicine, July 1992, Vol 152, pp 1371-72

Chapter 10

The Truth About Cholesterol

If the vitamin C theory is correct, you may be wondering about the role of cholesterol in heart disease. As it turns out, there is a strong correlation between vitamin C and cholesterol. The total cholesterol level in the blood can be predicted based on how much vitamin C one ingests. The less vitamin C one takes, the higher his total cholesterol levels become. According to Dr. Thomas Levy:

> Much of what we are told by our most trusted authorities turns out to be the exact OPPOSITE of what is true and what should be heeded. "Avoid foods that are high in cholesterol." This is yet another example of thoroughly misguided advice from our so-called health authorities. **— Thomas E. Levy, M.D., J.D.**

A Cholesterol Primer

Far from being health destroyers, ordinary cholesterol molecules are miniature miracles of nature that are essential for good health. You will not feel well if your cholesterol is too low.   These large molecules carry fat soluble nutrients throughout the body and help the body rid itself of fat soluble toxins. Cholesterol comes in different sizes and densities, such as low-density lipoproteins (LDL) and high-density lipoproteins (HDL). Even LDL ("bad") cholesterol is important for good health and should not be reduced much under what is considered normal.

Its chemical structure makes cholesterol a steroid. The body uses cholesterol as a building block for many other steroids, including the sex hormones. Vitamin D is another steroid. Sunlight stimulates the body's production of vitamin D from the cholesterol just beneath the skin. Vitamin D has made the news recently because studies have consistently shown that this fat soluble vitamin has strong immune-enhancing and anti-cancer properties.

The astonishing globules of cholesterol enable fat soluble substances to travel through the salt water that is our bloodstream. For example, the fat soluble nutrients vitamin E and coenzyme Q10 are carried through the bloodstream on LDL cholesterol molecules.

According to authors Michael R. and Mary Dan Eades:

> Although most people think of it as being "fat in the blood," only seven percent of the body's cholesterol is found there. In fact, cholesterol is not really fat at all; it's a pearly-colored, waxy, solid alcohol that is soapy to the touch. The bulk of the cholesterol in your body; the other 93 percent, is located in every cell of the body, where its unique waxy, soapy consistency provides the cell membranes with their structural integrity and regulates the flow of nutrients into and waste products out of the cells.

In addition, among its other diverse and essential functions are these: Cholesterol is the building block from which your body makes several important hormones: the adrenal hormones (aldosterone, which helps regulate blood pressure, and also hydrocortisone, the body's natural steroid) and the sex hormones (estrogen and testosterone). If you don't have enough cholesterol, you won't make enough sex hormones. — ***Protein Power, Michael R. Eades, M.D. and Mary Dan Eades, M.D.***

Cholesterol as a Detoxifier

Before he became active in alternative medicine, Board-certified cardiologist Thomas E. Levy worked closely with mercury-free dentist Hal Huggins. Levy noticed that after his cardiovascular patients had the mercury amalgams removed from their teeth, their total cholesterol readings dropped significantly within days. This observation prompted a literature search, which eventually spawned Dr. Levy's interest in vitamin C.

Levy found extensive experimental evidence that one function of cholesterol is to cleanse the body of toxins. The higher the toxic load, the more cholesterol the body produces. The human body has developed several similar defense mechanisms against toxins. For example, the production of mucous in the lungs and sinuses is another way in which the human body expels unwelcome particles such as foreign proteins, viruses, and toxins. Vitamin C has a well-studied "anti-histamine" property; as vitamin C levels increase in the sinus tissues, mucous production decreases.

Levy's study of the medical literature regarding cholesterol intrigued him. Scientists had discovered another important fact about cholesterol that is little known among members of the medical profession:

> Cholesterol is the body's natural detoxification mechanism. High cholesterol levels develop in response to the presence of toxins; the toxins are neutralized by the cholesterol. When patients with high cholesterol levels (over 240 mg%) had their mercury amalgams and sources of dental infection removed, these levels usually dropped dramatically within a few days. When toxin levels have been minimized, most people's cholesterol levels will be between 160 and 220mg%. — *Optimal Nutrition for Optimal Health,* **Thomas E. Levy, M.D., J.D.**

One interpretation of this finding is that if your cholesterol is high, you have toxins in your body.

If the body creates more cholesterol to help fight toxins, what happens to heart patients with mercury fillings and other toxic loads who are taking the popular cholesterol-lowering drugs? Medical doctors, by treating the symptom (elevated cholesterol) and not the root cause (low vitamin C), are unintentionally interfering with the body's ability to heal itself. Heart patients on these drugs lose a natural defense mechanism — the ability to generate high cholesterol levels. Cardiologists probably assume that any degradation in their patients is merely the natural progression of the disease.

> The concept that cholesterol can inactivate or neutralize a wide variety of toxins is not new; researchers have identified cholesterol as an inactivator of multiple bacterial toxins. Other researchers have shown that elevation of serum cholesterol actually served as a marker for various toxic exposures. The toxicity of

pesticides reliably elevates the cholesterol levels of those exposed individuals and one researcher showed that dogs exposed to low levels of methyl mercury developed progressively higher levels of cholesterol in the blood over time. — *Optimal Nutrition for Optimal Health*, **Thomas Levy, M.D., J.D., pp 89-91**

Vindicating Cholesterol

Cholesterol is necessary for life itself and is involved with many bodily functions. The substance is so important that if one stops eating foods that contain cholesterol, a feedback mechanism stimulates the liver to make more cholesterol. Below is a concise list from Drs. Eades of the bodily functions that cholesterol is known to be required for.

· Cholesterol is a precursor of vitamin D in the skin. When exposed to sunlight, this precursor molecule is converted to its active form for use in the body.
· Cholesterol is the main component of bile acids, which aid in the digestion of foods, particularly fatty foods. Without cholesterol we could not absorb the essential fat-soluble vitamins A, D, E, and K from the foods we eat.
· Cholesterol is necessary for normal growth and development of the brain and nervous system.
· Cholesterol coats the nerves and makes the transmission of nerve impulses possible.
· Cholesterol gives skin its ability to shed water.
· Cholesterol is important for normal growth and repair of tissues, since every cell membrane and the organelles (the tiny structures inside the cells that carry out specific functions) within the cells are rich in cholesterol. For this reason, newborn animals feed on milk or other cholesterol-rich foods such as the yolks of

eggs, which are there to provide food for the developing bird or chick embryos.
· Cholesterol plays a major role in the transportation of triglycerides or blood fats through the circulatory system.

Drs. Michael R. and Mary Dan Eades conclude:

A quick review of this list should give a better idea of what cholesterol does and dispel any notion that it is a destroyer of health to be feared and avoided at all costs. Far from being a serial killer, cholesterol is absolutely essential for good health; without it you would die. Without cholesterol we would lose the strength and stability of our cells, rendering them much less resistant to invasion by infection and malignancy. In fact, a grave sign of serious illness, such as cancer development or crippling arthritis, is a falling cholesterol level. — ***Protein Power*, Michael R. Eades, M.D., and Mary Dan Eades, M.D., 1996**

Studies: Cholesterol Does Not Cause Heart Disease

The International Network of Cholesterol Skeptics, a steadily growing group of scientists, physicians, and other academicians and science writers from various countries, is questioning the common dogma that dietary saturated fat and cholesterol cause heart disease. Not only is there no proof to support this hypothesis, says spokesman Dr. Uffe Ravnskov of Lund, Sweden, but the available scientific evidence clearly contradicts this claim.

Ravnskov says that the accumulated evidence from nine dietary trials presented by a group of British researchers in the 31 March 2001 issue of the *British Medical Journal* showed that not a single life was saved by dietary changes to reduce fat intake that went far beyond the official recommendations.

Dr. Walter Willet, Chairman of the Department of Nutrition at the Harvard School of Public Health, is the spokesman for the ongoing, longest running, most comprehensive diet and health study ever performed that involves nearly 300,000 subjects. As he recently reported, data from this study clearly contradicts the "low fat is good" health message and "the idea that all fat is bad for you; the exclusive focus on adverse effects of fat may have contributed to the obesity epidemic."

Hidden Dangers of Cholesterol-Lowering (Statin) Drugs

There is little evidence that lowering cholesterol protects humans from heart disease. According to the Life Extension Foundation, of the heart attacks that occurred in individuals under the age of 50, more than half occurred without any recognized risk factors.

According to noted nutrition expert Earl Mindell in his recent book, *Prescription Alternatives*, with Virginia Hopkins:

> There is absolutely no evidence anywhere that normal cholesterol floating around in the blood does any harm. In fact, cholesterol is the building block for all your steroid hormones, which includes all the sex hormones and the cortisones.
>
> Even slightly low levels of cholesterol are associated with depression, suicide, and lung cancer in older women... For most people, eating high cholesterol foods does not raise cholesterol.
>
> While a cholesterol-lowering drug will usually do a very good job of lowering your cholesterol, there's scant, if any, evidence that it will help you live longer or reduce

your risk of heart attack. If the American public had even a clue of how destructive these drugs are, they wouldn't touch them... Every information sheet on the most commonly prescribed cholesterol-lowering drugs will tell you that they cause cancer in rodents when taken long-term in relatively normal doses. It's also well-known that they can cause severe emotional imbalances in men along with a wide array of life-threatening side effects... The wisest course of action is to avoid these drugs.... — *Prescription Alternatives*, **Earl Mindell**

Cholesterol-Lowering Drugs - A Death Sentence?

We are now in a position to witness the unfolding of the greatest medical tragedy of all time – never before in history has the medical establishment knowingly created a life threatening nutrient deficiency in millions of otherwise healthy people. — **Peter H. Langsjoen, M.D.**

Ubiquinone (coenzyme Q10) is a vitamin-like molecule that our muscles require, especially the heart muscle, in order to do work. Many elderly individuals have low CoQ10 levels with associated fatigue, muscle aches, and possible heart failure. CoQ10 has become a popular heart medication in Japan, and as recently as 2001 it was only available in that country by prescription.

CoQ10 is not a vitamin for the young. Healthy humans under the age of 40 make about 500 mg of CoQ10 in their cells every day. The top selling statin cholesterol-lowering drugs such as Lipitor® and Zocor® interfere with the body's synthesis of CoQ10, and thus effectively lower circulating CoQ10 blood levels and also levels in the tissues. These drugs earn their makers in excess of $20 billion per year and are often touted as "life saving" by cardiologists and in the media.

According to CoQ10 expert and researcher Peter H. Langsjoen, M.D., the shocking fact is that statin drugs cause muscle damage that leads to heart failure:

> Lipid lowering drugs inhibit the formation of cholesterol by the liver. This mechanism also has other unwanted effects. The same enzymes that are involved in the production of cholesterol are also required for the production of an essential compound called coenzyme Q10; not surprisingly, lower cholesterol levels in statin users are accompanied by reduced levels of CoQ10.
>
> Coenzyme Q10 – also called ubiquinone, which means "occurring everywhere" – plays an important role in the manufacture of ATP, the fuel that runs cellular processes. Although it is present in every cell in the body, it is especially concentrated in the very active cells of the heart. Depriving the heart of CoQ10 is like removing a spark plug from your engine – it just won't work. Low levels of CoQ10 are implicated in virtually all cardiovascular diseases, including angina, hypertension, cardiomyopathy and congestive heart failure.
>
> In 1990, the drug manufacturer Merck sought and received a patent for Mevacor and other statin drugs formulated with up to 1,000 mg of coenzyme Q10 to prevent or alleviate cardiomyopathy, a serious condition that can cause congestive heart failure. However, Merck has not brought these combination products to market, nor has this drug company educated physicians on the importance of supplementing CoQ10 to offset the dangers to the heart from these drugs. Because they hold the patent, other drug companies are prevented from coming out with a statin/CoQ10 product.

In the last 15 years (roughly the time that statins have been on the market), the incidence of congestive heart failure has tripled. – **Peter H. Langsjoen, M.D.**

Julian Whitaker, one of the nation's leading alternative medical doctors, recommends that statin users supplement with at least 150 mg of CoQ10.

Statin Cholesterol-Lowering Drugs Linked to Cataracts

Richard Cenedella, Ph.D., held a press conference recently to announce that a recent multinational scientific study supports his hypothesis that brief exposure to high levels of some statin drugs may irreversibly damage the lens of the eye. The study by Christopher Meier of the University Hospital of Basel, Switzerland, suggests that taking the antibiotic erythromycin in combination with the popular statin cholesterol-lowering drugs may increase the risk of cataracts. Erythromycin slows removal of statins from the body and can greatly increase blood levels of these drugs. The study results published in *Archives of Internal Medicine* showed that a single course of an antibiotic, typically 10 days, appeared to double the risk of cataracts, and two or more courses tripled the risk.

Cholesterol-Lowering Drugs Increase Cancer Risk

All members of the two most popular classes of lipid-lowering drugs (the fibrates and the statins) cause cancer in rodents, in some cases at levels of animal exposure close to those prescribed to humans. This information is carried in the fine print accompanying the drugs and was recently published by the *Journal of the American Medical Association* (JAMA) on-line.

Would You Purchase This Supplement?

Most people would jump at the chance to purchase a dietary supplement that was capable of the following:

- Increasing libido (because it is a precursor to testosterone and other anabolic steroids and hormones);
- Helping to build strong bones (because it is a precursor to vitamin D);
- Providing protection from infections in the winter (because it is a precursor to vitamin D);
- Reducing the risk of cancer (because it is a precursor to vitamin D);
- Improving memory and aiding sleep (because it actively transports fat-soluble vitamins and coenzyme Q10 through the bloodstream and promotes the health and structural integrity of cells, especially cells of the brain and nervous system);
- Helping the body to expel foreign substances because of its natural anti-toxin properties; and
- Helping to extend life.

If you guessed this dietary supplement to be cholesterol, you are correct! Instead, chances are that you take a drug to lower circulating levels of this vital substance. The cholesterol-lowering drugs have been prescribed based on countless pharmaceutical company-inspired "studies." The drug companies keep running studies hoping to get the expected results, to no avail. No study that has released its unedited data has shown that heart patients reverse their disease by taking statins. This is not for lack of trying,

173

however, as according to television commercials Pfizer has now sponsored more than 400 studies of just one of these drugs.

Trusted authorities have told your doctor that lowering cholesterol is important for heart patients, and your doctor believes it. Cardiologists prescribe cholesterol-lowering drugs not because of a sound theory but because they are under the totally misguided assumption that these popular drugs will somehow reduce your chances of suffering a heart attack or stroke. The inescapable fact is that cholesterol-lowering drugs *increase* Lp(a).

Zetia® and Vytorin®: The Anatomy of a Debacle

In a press release on January 15, 2008, Merck and Schering said that not only did Zetia® fail to slow the accumulation of fatty plaque in the arteries but it actually seemed to contribute to plaque formation, albeit by such a small amount that the finding could have been a result of chance. Dr. Steven E. Nissen, the Chairman of Cardiology at the Cleveland Clinic, said the results were "shocking."

Cardiologists at the Cleveland Clinic said that a clinical trial of a widely-used cholesterol drug is raising questions about the medicine's effectiveness and about the negative behavior of the pharmaceutical companies that conducted the study. The popular cholesterol drug Zetia® was studied, and two years after the study ended, Merck and Schering-Plough, the makers of the drug, had still not released the results. Finally, the study's results were released under pressure from Congress, and they were not good. Not only did the drug fail to benefit patients in the two-year clinical trial, but the same drug that many patients take in a pill that contains it, Vytorin®, increased the buildup of harmful plaque in arteries.

The drug companies blamed the complexity of the data for the near two-year delay. Merck and Schering repeatedly missed their own deadlines for reporting the results, which led cardiologists around the world to wonder. At the same time, millions of patients kept taking Zetia® and Vytorin®. In January 2008, barely a month after news articles noted the delay and Congress pressured the companies to disclose the study's findings, the companies relented.

According to the press release, "This drug doesn't work. Period. It just does not work," said Steven Nissen, Chairman of Cardiology at the Cleveland Clinic. Michigan Congressman Bart Stupak, leading a Congressional investigation of the study, said, "It is easy to conclude that Merck and Schering-Plough intentionally sought to delay the release of this data."

Zetia® and Vytorin® (a combination pill of Zetia® and Zocor®) are prescribed 100,000 times a day and together generate annual sales for the pharmaceutical industry of at least $5 billion. Merck and Schering split the profits from the medicines equally, and according to *Forbes,* the drugs have been growth engines, with sales up by 25 percent in the most recent quarter over the same period last year.

Vitamin C is Nature's Perfect Statin

The hot-selling statin prescription drugs such as Lipitor® and Zocor® lower serum cholesterol by blocking a particular enzyme. The technical name for the enzyme is HMG-CoA reductase. The enzyme is used by the liver in its production of cholesterol. The first statin drugs to block this enzyme were released in the United States in 1987. The manufacturing and

175

promotion of these drugs has grown into a huge business; statins have become the most widely-prescribed class of drugs in history.

There are scientific experiments which clearly demonstrate that vitamin C regulates serum cholesterol. Linus Pauling was impressed by the results of an extensive series of experiments conducted during the 1970s by the Research Institute of Human Nutrition, Comenius University College of Physical Education, Bratislava, Czechoslovakia. Doctor Emil Ginter of this Institute ran extensive studies in guinea pigs and humans which showed that taking more vitamin C reliably lowers elevated cholesterol.

These researchers found that the effect of vitamin C on cholesterol is predictable. Ginter showed that the higher the total cholesterol level above 180 mg/dl, the more it can be lowered by taking vitamin C. Cholesterol levels normalized in the Ginter studies to 180 mg/dl.

In 1985, Harwood, et al. discovered that vitamin C is a so-called natural HMG-CoA reductase inhibitor, which means that ascorbic acid has the same cholesterol lowering properties exhibited by the statin cholesterol-lowering drugs, but without side effects. What Harwood discovered is a straight-line relationship between vitamin C concentrations and the inhibition of HMG-CoA reductase. We can extrapolate that similar to its antihistamine property, vitamin C has an anti-cholesterol property. When vitamin C levels are low, the body compensates and manufactures more cholesterol. When vitamin C levels in the cells that make cholesterol become high, vitamin C inhibits HMG-CoA reductase, lowering cholesterol naturally.

The problem with the artificial statin cholesterol-lowering drugs has been identified. These drugs concurrently reduce circulating levels of the ubiquinone coenzyme Q10 while raising circulating levels of Lp(a). A decline in CoQ10 – attributed to artificial statin drugs – has been linked to muscle pain, muscle myopathy, and congestive heart failure. Elevated Lp(a) increases the probability of a heart attack or stroke by 70 percent.

Vitamin C at dosages starting around 6,000 mg lowers total cholesterol in heart patients. These dosages probably lower Lp(a), and they help stimulate the body's production of CoQ10. Yet rather than promote vitamin C as a potentially safe way to manage cholesterol, drug companies invented statin drugs. One has to wonder, "What did the drug companies know, and when did they know it?" "Could vitamin C have been used as the model for the statin cholesterol-lowering drugs and the drug companies simply forgot to mention it?"

When my co-author's fiancé, William Cook of Louisville, Kentucky, suffered two heart attacks and triple coronary artery bypass graft surgery, he elected natural cholesterol reduction. Declining medical advice to begin a cholesterol-lowering statin drug, Cook instead began taking a commercial Pauling therapy formula in mid April of 2002. In eight months his total cholesterol dropped from 246 to 164 mg/dl and his LDL dropped from 163 to 101, all on extra vitamin C and Tower's *Heart Technology* ™. Cook's primary care physician commented, "I've never seen such a drop in cholesterol numbers. What have you done to get your numbers like this? If cardiologists knew about this they would be recommending this to all their patients."

Some have argued that for economic reasons we shouldn't trust the claims made about vitamin C. Vitamin C is the top-selling nutritional supplement, earning its makers roughly $180 million per annum. However, this pales in comparison to the top-selling cholesterol drugs. Statin drugs are sold worldwide to 25 million and earn their makers more than $20 billion annually.

Marcia Angell, former editor of the *New England Journal of Medicine*, in her book, *The Truth About The Drug Companies*, stated:

> The most startling fact about 2002 is that the combined profits for the ten drug companies in the Fortune 500 ($35.9 billion) were more than the profits for all the other 490 businesses put together ($33.7 billion). When I say this is a profitable industry, I mean really profitable. It is difficult to conceive of how awash in money big pharma is.

"DALLAS, TX – Nov. 10, 1998 – Stents reduce restenosis rates significantly over the past decade, but cardiologists have been somewhat frustrated by the problem of in-stent restenosis. Now researchers report at the 71st Scientific Sessions of the American Heart Association in Dallas, TX, say radiation therapy may produce a dramatic reversal of that emergent problem."

Chapter 11

Cardiology Revisited: Radioactive Stents

The ignorance of modern cardiology about the disease that the profession is charged with treating has resulted in a frightening development. Medical researchers are now headed down a bizarre path that subjects trusting heart patients to high-dose radiation pellets in their arteries!

BARCELONA — High-dose beta radiation, delivered along with balloon angioplasty or stenting, appears to reduce restenosis at six months in coronary vessels, Swiss researchers report. Using a system still under development, doctors at University Hospital in Geneva, along with colleagues in four other European centers, tested the feasibility of using intracoronary beta radiation in the treatment of de novo lesions. They found beta radiation produced "a significant, dose-dependent inhibitory effect on restenosis after PTCA (percutaneous transluminal coronary angioplasty) and a beneficial effect on remodeling," principal investigator Dr. Vitali Verin told doctors.

Conventional treatments and surgery for heart disease have serious after-effects. Restenosis (plaque regrowth) is common after heart surgery and angioplasty. The American Heart Association estimates that the coronary arteries reocclude after surgery in 40 percent of patients under the care of cardiologists. Bare metal stents did not solve the problem, so new, medicated, drug-eluting stents (DES) were invented to stop cell growth of the intima, the tissue on the inside of the artery, in a vain attempt to stop restenosis.

An even newer, and scarier, "high tech" way around this problem is the use of dose-dependent radiation to interfere with or perhaps destroy the ability of a patient's arterial tissues to heal.

This madness began in 1997, according to the European Society of Cardiology [Vol. 35, No. 33, October 5, 1999], with the work of the Scripps Clinic and Research Foundation in La Jolla, California. Heart doctors at this clinic reduced restenosis rates by using gamma radiation to stop neointimal hyperplasia, a common response to injury of the vessel wall during angioplasty.

Several FDA-sanctioned studies are now actively investigating whether intra-arterial radiation (in the form of pellets or "seeds") improves the success rate of coronary bypass surgery and angioplasty. We can surmise from the increasing level of mass desensitization (television news "reports") that this dangerous procedure is close to being sanctioned by the FDA.

PRESS RELEASE — Guidant's intravascular radiotherapy system, developed by the company's Vascular Intervention Group, is used alongside interventional cardiology procedures that help clear blocked arteries in the heart. Preliminary clinical data have indicated that the application of radiation to an artery in conjunction with interventional treatments has the potential to subsequently reduce or eliminate restenosis caused by smooth muscle cell proliferation. Restenosis, the re-narrowing of an artery, remains a major clinical challenge in interventional cardiology today.

If you find this alarming, it gets worse. Linus Pauling used the word "cure" in 1994 in describing the new therapy for heart disease described in this book. According to Pauling, the lesions that lead to heart disease are caused by a nutritional deficiency in humans – too little vitamin C. Linus Pauling's theory explains atherosclerosis and restenosis as the body's natural healing response to the damage caused by the vitamin deficiency.

Cardiology as a profession is silent on Pauling's theory and therapy. The same profession that shows not the slightest interest in studying a claim uttered by one of the world's preeminent scientists, a claim based on a Nobel Prize in medicine and scientific proof in animals, is more than willing to investigate the effects of high-dose radiation in patients whose stubborn arteries keep trying to heal themselves.

PRESS RELEASE — ...radiotherapy system is designed for patient safety and procedure simplicity and consists of a source wire, a source delivery unit and a centering catheter. The source wire is flexible and incorporates a radioactive isotope into its tip. The delivery unit stores the source wire when it is not in use and automatically advances and retracts it during a procedure. The source wire is advanced through the centering catheter, which is placed across the area to be treated. Extensive animal studies of...intravascular radiotherapy system, conducted

at Baylor, have demonstrated a significant reduction in proliferation after balloon angioplasty and stenting.

Has modern medicine lost its collective mind? Eureka! We can save the patient; we just take away his natural ability to heal! An argument could be made that medicine has mutated itself into the antithesis of healing - *by design*! It is incomprehensible that any doctor sworn to "do no harm" would willingly interfere with the body's natural healing process. With the knowledge of the Pauling/Rath theory, it is obvious that high-dose intra-arterial radiation treatments are wrongheaded, dangerous, inhumane, and cruel and nullify any chance for a nutritional-based therapy to succeed. The theories of Linus Pauling should certainly be investigated thoroughly before heart patients are subjected to radiation! Cardiologists seem all too willing to subject heart patients to a treatment that is cruel and unusual.

This begs the questions, "Why is cardiology so far off track? " "Do cardiologists have eyes? " "Can they not see what works and what does not work?"

Pauling used the word "miraculous" and he was not overstating the effect. Since 1995 we have personally observed end-stage heart patient after end-stage heart patient seemingly "cured" within weeks on the high-dose Pauling therapy.

By "cured" we mean that end-stage cardiovascular patients report that angina pain ceases completely, color returns, blood pressure drops, blood flow increases, blockages disappear, heart rates drop, lipid profiles normalize, energy increases, and the sense of well-being increases. Patients who previously failed treadmill stress tests now report passing without surgery or any other medical intervention. Patients who were hardly able to walk before adopting the Pauling therapy report being able to dig fence post holes and cut down trees within months. Doctors have even told patients that "new blood vessels have grown that are now feeding the heart" to explain the increased blood flow.

Based on these observations and similar reports, we consider the proposed use of radiation to be unethical. We hope the FDA will not approve the procedure, making it criminal. To the extent that intra-arterial radiation treatments interfere with the ability of the arteries to heal themselves normally (and they do), these radiation treatments nullify almost any chance of success from the Pauling therapy protocol.

We sincerely hope that the intra-arterial radiation research is based on ignorance. Since radiation is being widely studied, it is evident that cardiovascular doctors as a group are entirely ignorant of the Pauling *Unified Theory*. But again, "Why?" "Why are the most educated professionals the most ignorant, or stubborn, about the condition they are licensed to treat?"

No profit-oriented drug company will inform cardiologists about the Pauling/Rath *Unified Theory*. So, then, "How does the average nuclear cardiologist learn about it?" Answer: "He doesn't." To our knowledge, no cardiologist or heart surgeon has ever been informed about vitamin C and lysine from a respected authority, except for the one group who heard of the theory from our pharmacist friend, Dan, as related in Chapter 1.

It would seem medical doctors do not realize that they could cure most patients in less than six months simply and safely, with the favorable effects becoming pronounced after a mere two weeks. Ignorance perpetuates itself. Cardiologists cannot believe this to be true or they would know about it.

Radiation: The Wrong Answer to the Wrong Question

Given that plaque formation is a surrogate healing process, doctors should not be surprised that plaque reoccurs after invasive surgery. Mainstream medical science has known since 1989 that only Lp(a) (not LDL) cholesterol binds to form atherosclerotic plaques. Pauling and Rath, quick to recognize the importance of this finding, identified Lp(a) as an evolutionary surrogate acting in the place of low vitamin C in humans.

Not only does intra-arterial radiation interfere with the healing of chronic scurvy, but the risk is unnecessary. Patients who are subjected to radiation will not heal properly and will most likely suffer more than if the stents they received were bare metal. Should the patient survive, how can anyone know the dangerous side effects that radiation may itself cause (e.g. cancer)?

The doctors who utilize radiation in an attempt to stop restenosis must not understand the disease they are attempting to treat. If they do understand, in my opinion the action of using a radioactive stent may constitute doing "harm." We conclude, therefore, that the true nature of heart disease is unknown to cardiologists who would otherwise know that restenosis is completely preventable and that high-dose radiation procedures are unnecessary and harmful.

Damaged arteries must heal. When a cardiothoracic surgeon performs a bypass procedure or otherwise damages an artery during angioplasty, he must not be surprised that the scabs (atherosclerotic plaques) form again. Any therapy that unnaturally interferes with this healing process (beta radiation) poses a grave risk to the overall health of the patient and there is no ethical basis for it.

The average person might assume that since vitamin C and lysine are harmless, cardiologists would have nothing to lose by trying the Pauling therapy. However, believing that their profession is based on strict science, cardiologists generally ignore unapproved therapies. These health professionals are taught that there is no value in nutritional therapies for heart disease, especially vitamin C.

Pauling even argued that a full clinical trial was not necessary before recommending vitamin C, with so many positive results and so little downside. If Pauling was wrong, little or no harm will have been done; if cardiology is wrong, millions will die needlessly.

In 1995 we didn't know whether Pauling was right or wrong, so we made it a point to ask the thousands of callers about their daily vitamin C intake. We noticed a familiar pattern. Callers who said they had severe, end-stage cardiovascular disease told us that they did not supplement or that they had stopped their supplemental vitamin C. We estimate that of the thousands of callers since 1995 who have reported that they had severe heart disease, 95 percent admitted to taking less than 500 mg of vitamin C daily.

I have stated publicly my postulate that 10,000 mg of vitamin C will generally protect one from CVD and asked that anyone with experience to the contrary contact me. I have yet to hear from any person without prior cardiovascular disease who has taken more than 10 g (10,000 mg) of vitamin C daily for at least one year who has any evidence of heart disease (and neither has Life Extension Foundation). Life Extension recently reported a published study of high-dose vitamin C users, none of whom had signs of atherosclerosis (see http://www.lef.org/featured-articles/may2000_vitamin_c_01.html).

Heart patients in hospitals are all vitamin C deficient. Cardiologists routinely tell their patients that there is no value in supplemental vitamin C above the RDA.

The results of our informal survey have been so overwhelmingly one-sided that we conclude that professional cardiologists are blinded to the facts by economic and/or political considerations. (*It is interesting to note that from our experience, retired cardiologists are not so blind.*)

The American Heart Association estimates that the cost of heart disease was $326 billion in the year 2000, including lost productivity at work, lost wages, etc. If medicine had acted on Pauling's claims back in 1994 or 1995, the investigation would be long over and we surmise that more than a million lives and trillions of dollars could have been saved. Ignoring the Pauling/Rath heart disease cure is not a crime that can be pinned on any one person or organization, but it is potentially the most costly suppression of a health cure. For more on how the suppression works, see The Pauling Therapy Handbook, Volume II.

Anyone connected with alternative medicine cannot help but become paranoid. That the lack of vitamin C is the key risk factor in heart disease has been shown by more than 650 studies, though these studies are being ignored by the medical journals, and hence by cardiology. The problem facing cardiology from a business standpoint is that the effects of the Pauling therapy are rapid, pronounced, dramatic, and seemingly without downside. Vitamin C and lysine are more effective than any medication in the arsenal of drugs available to cardiology, and it doesn't take heart patients long to judge whether the Pauling therapy works. Although hard to believe, perhaps there *is* an ulterior motive behind the ongoing research into intra-arterial radiated stents.

We can't help but think that maybe the hidden agenda behind the push to sanction radiation pellets is to keep heart patients perpetually ill. Now that knowledge of the Pauling cure is spreading rapidly thanks to the Internet, intra-arterial radiation may offer nuclear cardiologists a defense against impending financial ruin. Intentional or not, FDA-approved intra-arterial radiation would help to ensure the lucrative incomes of heart doctors. Cardiology cannot compete head-on with the Pauling and other alternative therapies.

The Pauling/Rath *Unified Theory* can help keep the dangers of intra-arterial radiation in perspective. The blood vessels of heart patients are not getting enough vitamin C to produce collagen and will continue to deteriorate. Plaques form and come to the rescue as Lp(a) begins the process of healing the artery. Simply removing the plaques by widening the artery via angioplasty without restoring the vein or artery to health is like tearing a scab off of a wound. Intra-arterial radiation compounds this basic theoretical problem. One should not remove the scab until after the tissue underneath has begun to heal. The body requires adequate vitamin C in order for veins and arteries to heal themselves. If radiation destroys the ability to heal, the doctor has created the worst of all possible scenarios.

Who cares about the patient? Apparently the cardiovascular patient's health takes a back seat to the health of the technician who applies the radiation! According to Dr. Spencer King, Professor of Medicine (cardiology) and Radiology at Emory University Medical Centre in Atlanta, "Beta radiation travels a short distance in the artery and does not leave the body, so it has a great attraction for the operator, because there's no radiation exposure."

If a doctor prescribes a treatment such as intra-arterial radiation that interferes the body's ability to heal, then no alternative or other complementary therapy (e.g. EDTA chelation) is likely to ever be effective. Any doctor who implants radioactive stents, even if these atrocities are someday approved by the FDA, would be well advised to read this book and especially this chapter. Ignorance should be no excuse. The patient should tell his doctor that he would first want to try the Pauling therapy. If the doctor refuses and insists on a radioactive approach, then the patient should find another doctor.

My advice: *Avoid radiation for restenosis as if your life depends on it.*

"American corporations have begun to realize that a huge chunk of North America's Gross National Product (GNP) is being funneled to FOREIGN DRUG companies - multinational drug syndicates based in France, Germany, etc. - through the grossly inflated price of pharmaceuticals in the United States. North America is being bled. And, big American corporations are picking up a big part of the tab in increased health costs for their employees. That means that US companies have to raise the price of their manufactured goods - which means they can't compete in world markets with the countries that host big pharma. "Big pharma" is NOT an American business - not at all. The only thing "American" about big pharma is its being half of the word "Anti-American." — Tim Bolen, Consumer Advocate

Chapter 12

Universal Health Care Vouchers

The treatment of chronic illness is a cash cow that has helped to create America's most profitable corporations. Sadly, the pharmaceutical industry, the hospital industry, and health care providers share a vested interest. They all make money from you when you are ill. This has created a perversion in the free market because vastly more money can be made by treating illness than by keeping the public healthy.

The current free market forces explain why Linus Pauling has been denigrated and why the Pauling/Rath vitamin C theory has been ignored in the media. But what if your doctor made more money by keeping you well than by treating some ailment? The new government voucher plan proposed here, *Universal Health Care Vouchers,* would reorient the free market to reward doctors who are able to keep their patients healthy.

The *Universal Health Care Vouchers* plan would allow us the choice of family physicians. Its viability is based on the fact that there are low-cost and effective cures such as the Pauling therapy that can replace the need for expensive, toxic, and dangerous prescription drugs.

The *Universal Health Care Vouchers* plan would work by paying the doctor even when his patients are well. This would not only correct the quirk in the free market but it could *save* the United States billions of dollars in health care costs. This means that because of the Pauling therapy and other natural therapies like it, we could literally have our cake and eat it, too. We could have universal health care and pay less for it than we pay for health care today.

One ramification of the success of the Pauling theory and therapy is the possibility that most if not all heart prescriptions could be eliminated. These drugs are dangerous, though their continued use seems inevitable. Most universal plans pay for and thus stimulate the use of these and other dangerous drugs.

There would be reason to expect similar savings from the reduction or elimination of prescription pharmaceuticals for other chronic diseases. The big losers from the *Vouchers* plan would be the pharmaceutical giants, health care providers, hospitals, and funeral homes.

This *Universal Health Care Vouchers* plan is designed to rescue medical doctors and allow them to profit even when their patients don't fall ill. Only the pharmaceutical companies whose products are judged useful by doctors would survive the shaking out process. The plan is similar, but not identical, to the *School Voucher* plan of the late Nobel Prize economist

Milton Friedman, whose vouchers give parents a choice of which school their children will attend.

Rewarding Health, Not Illness

When the free market rewards sickness, more sickness is to be expected. A study of inefficiencies and dangers in American hospitals conducted by the Institute of Medicine yielded alarming statistics. Included in their chilling findings were the following atrocities:

- Adverse reactions to medications kill 108,000 people every year.
- Unnecessary surgeries kill 7,000 people.
- Hospital medication errors kill 12,000 people.
- Other hospital errors kill 20,000 people.
- Hospital-contracted infections kill 80,000 people.

This hair-raising report, summarized in the Institute's July 2006 *Preventing Medication Errors*, also stated that millions more Americans are injured by the negligence of hospital personnel and by being given the wrong or inappropriate prescription drugs. The authors of the Institute of Medicine study considered these estimates to be conservative.

Health insurance pays for these disasters. The drug companies make so much money from a single class of prescription drug, rightly or wrongly dispensed, that they could fund the purchase of 40 professional sports teams each year (i.e. $20 billion annually from cholesterol-lowering statins)!

In the United States the cost of treating illness has been estimated to be between 7 percent and 12 percent of the Gross

Domestic Product (GDP). Consider the number of lives that could have been spared had the economic incentives been different and the Pauling therapy (non-drug) cure been explored and applied in general practice. We have no way of knowing how many hospital visits could have been avoided.

Instead, heart patients are routinely treated with the statin cholesterol drugs that deplete the body of endogenous coenzyme Q10. In the opinion of CoQ10 experts, statin cholesterol drugs that are given to heart patients are causing their patients' heart conditions to worsen.

The Appeal of Vouchers

The Federally-sponsored *Universal Health Care Vouchers* plan would change the current free market incentives. The doctor would receive his annual payment when a family gives him a voucher. A viable plan must enable a family physician to earn a good living, even if he isn't doing much work, as long as families are willing to give him the yearly vouchers. Under the plan, a doctor would be paid whether or not his patients ever became sick. Any patients that became ill would increase the doctor's expenses.

When the doctor accepts the voucher, a contract is created. The physician accepts the voucher in exchange for agreeing to provide the family with complete health care for one year. The physician assumes the responsibility for securing catastrophic group health insurance for his practice and in this way covers the cost of any patient trauma or necessary hospitalizations. Otherwise, the doctor provides the patient care.

Doctors would therefore earn more money under a voucher system by keeping their patients healthy and out of their offices and hospitals. The doctor would have incentive to provide proper service because of the possibility that the family could decide to give next year's voucher to another physician.

Vouchers would change the economic incentives and place the major health care decisions in the hands of doctors. We propose that this program should be Federally funded on the basis that it would substantially reduce the overall cost of health care to society. Because the treatment of any illness would now cost the doctor and not the patient, there would be strong economic incentive for doctors to prevent and minimize illness. The economic forces would motivate doctors to closely scrutinize any potential cures that might have been disregarded under the current, "perverted" free market system.

Perhaps the greatest implication would be the reduced use of prescription drugs. Prescription drugs are not only expensive but they are toxic, potentially very dangerous, and often lead to other illness. Today, the medical doctor relies on the monopoly that he has to write prescriptions to attract patients. Under the *Vouchers* plan a doctor would have much less incentive to write prescriptions and much more incentive to find safer and more effective natural remedies.

The Vouchers Plan

The Federal Government would provide health care vouchers to all legal residents/tax-paying families in the same manner as the proposed school vouchers. The family voucher could be electronic. It would be assigned to the physician of choice in exchange for complete family health care for one year.

Under the *Vouchers* plan, the family physician would become the focus of family health care. He would make the important medical decisions in exchange for the cash value of the voucher. An important requirement would be that the family physician be responsible for total health care. The doctor would either pay for patient hospitalizations out-of-pocket, charge patients the costs if he could and still attract vouchers, or elect to purchase group health insurance to cover catastrophes and necessary hospitalizations. Thus, every major economic and medical decision would be made by the most knowledgeable party — the family physician.

The value of the vouchers must provide incentive for physicians to behave differently. The value of the voucher must increase for each additional family member. Doctors must be able to earn a significant income without their patients having to become sick. Physicians must retain the right to refuse family vouchers.

The cumulative value of the vouchers should be 2.5 percent to 3 percent of the GDP, saving current costs by at least 1 percent of the GDP and possibly as much as 10 percent of the GDP.

The plan should pay the physicians for all patients whose vouchers they attract, regardless of whether or not the patients require a doctor's services. In order to change the economic incentives, the vouchers should reward physicians the most when patients remain healthy and do not require expensive medical attention.

Families should have the power to select their physicians each time the voucher is issued, probably yearly.

Medicare would remain and be the safety net for those who cannot find doctors to accept their vouchers.

The Impact of Vouchers

The key feature of this plan would be the government-induced change in free market forces that affect health care costs. The market forces are redirected away from rewarding sickness, and instead family physicians would earn more by keeping patients well, out of doctors' offices, and out of hospitals. Skilled physicians would be able to service more patients and realize increased net income by attracting more vouchers.

Consider our experience with the non-prescription, non-toxic Pauling therapy for cardiovascular disease. Today, most cardiologists would be afraid of this plan because their patients currently only seem to worsen. However, alternative doctors with knowledge of the Pauling therapy would readily accept vouchers, even from critically ill heart patients.

Such thought experiments help us to predict that the reorientation of market forces would bring the total cost of health care down from the current 7 to 12 percent of the GDP to less than 4 percent of the current GDP. Saving even 1 percent of the U. S. GDP would be an enormous boost to the U. S. economy.

There would be a savings from the elimination of regular health insurance claims, but the greatest savings would lie in the elimination of toximolecular prescription drugs that do more harm than good.

Today's Medicine Kills

According to the American Medical Association, even without the benefit of health insurance payments Americans make more than 100 million more visits to alternative practitioners than to orthodox medical doctors. As a nation we spend billions looking for symptom relief from toxic prescription drugs. The *Universal Health Care Vouchers* plan would change almost everything about health care. The invisible hand of the new economics would reward the family physician most when his patients remain healthy and out of the hospital. No longer would the free market incentives reward attempts at fixing people after they have been harmed. Under the new incentives outlined here, chronic diseases that are curable would be eliminated.

A Brilliant Atomic Scientist Has a Brilliant Suggestion

If our politicians decide that selling a *Universal Health Care Vouchers* plan would be impossible or too risky, here is a brilliant idea that achieves nearly the same effect. Theodore P. Jorgensen, Ph.D., is a national hero. He is a Harvard professor of physics, now retired. Doctor Jorgensen was a member of the Manhattan project during World War II, having worked on the atomic bomb at Los Alamos.

In 2003, Dr. Jorgensen, then 99 years of age, sent the following note. His idea would probably save the United States billions of dollars in prescription drugs and health care. All that has to happen is for *the United States Federal Government to provide ascorbic acid (vitamin C) free to all citizens.*

Dr. Jorgensen writes in private correspondence:

> ...for many years research using ascorbic acid was done using very small amounts of the substance. It took many years before it was discovered that ascorbic acid could be used to produce fabulous results when used correctly in medical and clinical research...
>
> It was discovered that most animals produce their own ascorbic acid and that human beings, apes,

monkeys, and guinea pigs could not make any at all. The conclusion of the thinking on this problem was that those animals which could not make ascorbic acid had a genetic defect involving one enzyme which was lost millions of years ago because ascorbic acid was so easy to obtain in the foods then available.

In order to obtain this amount of ascorbic acid a human being should have, work was done to find what other animals made for their own use. The result of this study put the value of ascorbic acid at 2.3 to 10 grams per 154 pound man in good health.

It is virtually impossible for any person to obtain this much ascorbic acid per day from ordinary or casual ways. This also indicates that human beings are living with dangerously low levels of ascorbic acid. The above information gives some idea of the reason our cost of health care is so high and our average age of death is so low. This problem is a National disgrace and should be attacked on a National basis. There are two reasons why this should be done. One reason is that a free supply of ascorbic acid to every person would lower the cost of health care in a major way. — **Theodore P. Jorgensen**

"Well, I don't know that there is a need for a randomized prospective, double blind controlled trial when you get evidence of this sort, the value of large intakes of vitamin C and also of lysine for preventing the deposition of atherosclerotic plaques, and preventing death from cardiovascular disease." — Linus Pauling, *Unified Theory* Lecture on Video

Chapter 13

A Theory is a Terrible Thing to Waste

The information presented in this book is based on more than 12 years of experience with the vitamin C theory of heart disease. The Pauling/Rath *Unified Theory* explains the cardiovascular disease process. The knowledge of this theory, most of which is available in a videotaped lecture given by Linus Pauling in 1992, has allowed thousands of heart patients to improve their lives, just as Pauling told us it would.

A good scientific theory brings order to chaos. The vitamin C *Unified Theory* filters the mass of accumulated scientific data and anecdotal evidence and bares the essentials. It makes sense out of seemingly contradictory experimental results. We can understand what Pauling understood — that good science rarely produces contradictory results. However, the data must be viewed in the correct light. For example, the vitamin C *Unified Theory* predicts that clinical trials using low doses of vitamin C will fail. Without the theory, we would have no means by which to filter studies that utilized improperly low dosages.

Data that would invalidate the theory must be the most closely scrutinized. If contradictory data were found to be valid, the data would threaten the theory itself. So far, in the case of vitamin C, studies with relevant dosages show much larger benefits than studies with smaller dosages. Most studies run by medicine have generally been run with amounts of the vitamin not much above what is available in food. However, should a trial of the large dosages recommended by Pauling repeatedly fail in the investigation of heart disease, those results would have invalidated the Pauling/Rath theory, or at least would have prompted a major adjustment. Unfortunately, no trials of large doses of vitamin C together with lysine as recommended by Pauling have been published. At present we must rely on the thousands of anecdotal reports provided by those taking large amounts of vitamin C and lysine.

A tenet of the vitamin C *Unified Theory* is that during evolution our ancestors lost the capacity to make vitamin C in the liver. For billions of years our predecessors made their own vitamin C, but Pauling and others estimate that our ancestors lost the ability to make ascorbate anywhere between 3 and 30 million years ago. Part of the reasoning behind this timeframe is that several high order primates share man's inability to manufacture vitamin C, and he may have had a common ancestor who started passing the GLO mutation to his children. This negative trait has been carried down until today. The lack of vitamin C in the bloodstream is not because we do not ingest enough vitamin C. Rather, it is because of a genetic defect in human DNA. There is little evidence that any of the species on this planet evolved to eat the amount of vitamin C necessary to attain the best of health, and we humans are no exception. Vitamin C is an atypical vitamin, and that which appears in food bears little relation to the amount we require.

Dietitians and orthodox nutritionists often make the fallacious argument that we should only consume the amount of vitamin C available in the diet. They point out that we would have to eat "bushels of oranges" to obtain the amount of vitamin C that Linus Pauling recommended. The issue is not what we eat but how we recreate vitamin C in the bloodstream in the amounts that would otherwise be there if the so-called GLO genetic defect had never occurred. Without genetic engineering, these high, sustained amounts of ascorbate in the blood and tissues can only be achieved by taking vitamin C throughout the day in amounts totaling at least 6,000 mg per day. Anyone who tells you differently does not know what he is talking about.

Linus Pauling himself ingested 18,000 mg daily because he computed that an animal of his body weight, on average, would produce 9,000 mg of ascorbate daily. Pauling also knew that as much as half of the vitamin C we ingest is lost, so he doubled his intake.

Owen's Epilogue

My paternal ancestry may be considered a microcosm for the *Unified Theory*. The theory holds that vitamin C levels in the blood beneath optimal levels inevitably lead to cardiovascular disease in otherwise healthy people.

My father was from a large family with a history of heart disease. With the exception of one surviving sister, his nine brothers and sisters all died of cardiovascular disease. The incidence of heart disease and fatal heart attack on my father's side of the family correlates to known vitamin C intake.

My father Milton was the youngest. He didn't smoke. He didn't consume alcohol. He tried to remain active and he took daily walks. He rarely became ill and almost never caught a cold. However, he had an extremely low tolerance for vitamin C. He would experience diarrhea if he took more than 200 mg of vitamin C in any one day, and so he did not take vitamin C supplements. The late doctor Robert Cathcart would have considered my father abnormal because of his own experience with more than 20,000 patients. Cathcart found that almost everyone can tolerate at least 4,000 mg (4 grams) of vitamin C daily in divided dosages. My father had what turned out to be a fatal heart attack at the age of 69 while visiting my mother in the hospital and died a few days later in February 1989 during an angioplasty procedure.

My father's only surviving sister is now 94 years of age and is in reasonably good health. My aunt agreed to read Linus Pauling's book, *How To Live Longer and Feel Better,* when she was in her late seventies. My aunt tells me that since reading Pauling's book she has made sure to get her daily vitamin C, usually in orange juice.

One older cousin, also from my father's side of the family, was diagnosed with cardiovascular disease several years ago. He underwent an angioplasty procedure and continues to be under the care of a cardiologist. He avoids fats in his diet and takes a statin cholesterol-lowering drug. He also has difficulty sleeping. Although my cousin is of the firm belief that I am off base in recommending so much vitamin C, he has been prudent enough, as insurance, to take an array of nutrients similar to the protocol outlined in Chapter 7 featuring vitamin C and CoQ10. My cousin does not find it strange that he is the only patient his cardiologist has "graduated" to wellness.

My older brother has a history of heart disease and has suffered three heart attacks and coronary artery bypass grafting. He has been a longtime vitamin C and supplement user, though like our father he has a low vitamin C bowel tolerance and can barely tolerate 4,000 mg of ordinary vitamin C daily. My brother's intake was apparently not enough to avoid heart disease, especially in someone who smoked cigarettes for most of his life. Since the last cardiac episode my brother has finally quit smoking and has also begun to take Lypo-Spheric™ vitamin C from livonlabs.com.

Lypo-Spheric™ C is a special form of vitamin C that does not cause gas or diarrhea. It should be considered by any person with a low vitamin C bowel tolerance. The vitamin C in Lypo-Spheric™ C is encapsulated in microscopic particles called liposomes. These nano-sized particles preserve the C and are well absorbed in the intestinal tract. Recent test results reported by Drs. Hickey and Roberts have shown that Lypo-Spheric™ C increases blood levels of vitamin C. Therefore, its use is indicated in persons who cannot tolerate more than 5,000 mg of ordinary vitamin C daily without gas or diarrhea. Lypo-Spheric™ C should be taken in addition to ordinary vitamin C as one approaches bowel tolerance.

Children Require Vitamin C, Too

The biochemist Irwin Stone first interested Linus Pauling in vitamin C research. Dr. Stone believed that most humans are born in a vitamin C-deficient or scorbutic state and that this is the primary reason that Sudden Infant Death Syndrome (SIDS) is prevalent among newborns. Proper nutrition is of utmost importance for an expectant mother. Perhaps the best thing she can do for herself and her future child is to take high doses of oral vitamin C before, during, and after her pregnancy. The tissues are saturated with vitamin C during the pregnancies of all other species. Parents should read Pauling's book, *How to Live Longer and Feel Better*, and supplement their children according to the entire Pauling regimen described in Chapter 1. If one doubts Pauling because his vitamin A recommendation seems too high, he should consider that the World Health Organization saves the lives of thousands of children in third world countries with single injections of 200,000 to 500,000 IU of vitamin A. In the author's estimation, not only are Pauling's recommendations entirely safe for children but failure to give children that which Linus Pauling recommended is unsafe. If your child is getting the vitamin C, vitamin A, and vitamin D3 that he or she requires, it is unlikely that your child will ever become noticeably ill.

My youngest son has had optimal levels of vitamin C in his blood since before he was born, in that my wife took 6,000 to 8,000 mg of vitamin C daily both before and during her pregnancy (which is even more than the 3,000 to 5,000 she usually takes). My wife took even more oral ascorbate during lactation. She was healthy during her pregnancy and my son has been the picture of health since the day of his birth. We have generally followed Linus Pauling's advice and my son, now 18 years of age, has been on the Pauling regimen throughout his entire life. He is never sick and he hardly ever feels sick.

Our experience with vitamin C supplementation has taught us that we rarely have to feel ill. When vitamin C tissue levels are high we may still contract a viral or bacterial infection, though rarely do we suffer the symptoms that were common before we started taking vitamin C. We now know that feeling sick only occurs when the tissues have become depleted of the nutrients needed to fight the infection, most especially ascorbate. If vitamin C is given in large doses, especially at the onset of an illness, one still contracts the infection but does not feel ill. Other than one day in grammar school, my son has not missed a single day of school because of illness. He had perfect attendance in high school. He is not an athlete, yet during physical fitness testing he was routinely at the top of his class, doing the maximum number of sit-ups and push-ups.

I have personally been taking large amounts of vitamin C, mostly as ascorbic acid, for more than 20 years. In 1983 I lost a bet with my mother and reluctantly read Pauling's first book about vitamin C. The entire family was getting sick and I criticized my mother for popping so many 1,000-mg vitamin C pills. At that time I was of the mindset that too much of anything, even water, was risky. The bet I had made with my mother was that if she didn't become sick, I would agree to read a book. With the exception of my mother, everyone in the household — my father, my brother, my wife and I — became quite ill and had the same hacking cough that lasted for weeks. My mother never contracted the illness, and so I read the Pauling book.

Prior to reading *Vitamin C, The Common Cold and the Flu*, I had no interest in vitamins or nutrition. After reading Pauling's book I began to take 3 grams of vitamin C daily. Soon thereafter multiple recurring infections and illnesses were "cured," never to return. Three years later, after reading Pauling's new book that was published in 1986, *How to Live Longer and Feel Better*, as well as Dr. Robert Cathcart's famous paper, *Titrating Vitamin C to Bowel Tolerance* (which Linus Pauling had kindly mailed to me in response to a thank you letter), I increased my daily vitamin C to my bowel tolerance, from 3,000 mg to 18,000 mg (18 g), which coincidentally matched Pauling's intake. With the consistent daily intake of 18 to 20 grams of vitamin C, several remaining minor medical conditions that had persisted at only 3,000 mg (3 g) daily subsequently cleared up, such as hay fever and gouty arthritis. Those conditions have never returned.

Fascinated by the rapid betterment of my own health and wondering why vitamin C was being ignored by the medical profession, I began intensive research into vitamin C and the so-called orthomolecular nutrition that continues to this day. This book is a compilation of my own articles that have been published in various alternative medicine publications, articles written and edited with the help of my mother. My mom was the first person in our family to research nutrition after she was diagnosed with rheumatoid arthritis subsequent to a serious viral infection at the age of 40 years.

I have followed Linus Pauling's regimen now for more than 20 years. My total cholesterol, usually 180 mg/dl, dropped to 160 mg/dl after I began to add lysine to my daily regimen. I do not know what my Lp(a) was before I began to take vitamin C at high doses, though currently my Lp(a) value is below 5 mg/dl. I would still contract viral infections in the winter despite my high vitamin C intake until two years ago when I discovered the magic of vitamin D3. I have since added this vitamin to Pauling's regimen in the fall and winter, taking 1,000 IU all year but 2,000 IU to 5,000 IU during the winter, and I have enjoyed the past two winters without so much as a sniffle.

At the age of 50 I began lifting weights again and working out with my then 16-year-old son on a Bowflex exercise machine. It was necessary to add only one supplement - 200 mg of coenzyme Q10. My son and I have now been working out several times per week for more than two years without aches or pains.

The Great Suppression - The Pauling Therapy Handbook, Volume II

We could not fit everything we know about this story into one book. The reasons why the vitamin C *Unified Theory* has never been tested by medical science are sad, complex, disturbing, and almost unthinkable. The Pauling Therapy Handbook, Volume II will detail our growing understanding of how those orchestrating the suppression of medical cures have been able to accomplish such a horrendous feat. If this book made you cry, the next one will make you weep.

Volume II will also include the story of the 100-year-old, "100-percent effective" cure for cancer.

Resources

For a referral to an alternative medicine health care professional in your area:

http://www.acam.org
American College for Advancement in Medicine
24411 Ridge Route, Suite 115
Laguna Hills, CA 92653
Phone: (949) 309-3520

For Information on Orthomolecular Medicine:

http://www.orthomed.org
Journal of Orthomolecular Medicine
Published by the International Society of Orthomolecular Medicine
16 Florence Avenue
Toronto, Ontario, Canada M2N 1E9
Phone: (416) 733-2117

http://www.OrthomolecularVitaminCentre.com
Orthomolecular Vitamin Information Centre, Inc.
Abram Hoffer, Ph.D., R.N.C.P.
Suite 3A-2727 Quadra Street
Victoria, British Columbia, Canada V8T 4E5
Phone: (250) 386-8756

To Order Lp(a) Testing:

http://www.atherotech.com
Atherotech, Inc.

201 London Parkway
Birmingham, AL 35211
Phone: (800) 719-9807

http://www.lef.org
Life Extension Foundation
P.O. Box 407189
Ft. Lauderdale, FL 33340-7198
Phone: (954) 766-8433
Orders: (800) 544-4440
Advisors: (800) 226-2370

To Order the Linus Pauling *Unified Theory* Lecture on DVD:

http://www.PaulingTherapy.com
Intelisoft Multimedia, Inc.
P.O. Box 73172
Houston, TX 77273
Phone: (800) 894-9025
Outside the USA: (281) 443-3634

http://www.TowerLaboratories.com
Tower Orthomolecular Labs
3432 Bruce Street, Suite 3
North Las Vegas, Nevada 89030
Orders: (877) 869-3752

http://www.lef.org
Life Extension Foundation
Phone: (800) 544-4440

To Order Pauling-therapy® Nutritional Products:

http://www.Cardio-C.com
Pauling Therapy Formulas, Inc.
P.O. Box 3097
Lisle, IL 60532
Phone: (800) 894-9025

http://www.PaulingFormulas.com
Pauling Therapy Formulas, Inc.
Phone: (800) 894-9025

http://www.HeartTechnology.com
Tower Orthomolecular Labs
Orders and Product Support: (877) 869-3752
Shipping and Billing Questions: (702) 876-5805

http://www.HeartDiseaseCauseAndCure.com
Sally Snyder Jewell
Pauling Therapy Enterprises
Toll Free: (877) 753-9355

For More Information on the Pauling-therapy® and to Read More Testimonials:

http://www.PracticingMedicineWithoutALicense.com
Owen R. Fonorow

http://www.PaulingTherapyForHeartDisease.com
Sally Snyder Jewell
Pauling Therapy Enterprises
Toll Free: (877) 753-9355

http://www.PaulingTherapy.com
Intelisoft Multimedia, Inc.

http://www.PaulingTherapyFormulas.com
Pauling Therapy Formulas, Inc.

http://www.TheCureForHeartDisease.com
Owen R. Fonorow

For the World's Most Accurate and Up-to-Date Information on Vitamin C:

http://www.VitaminCFoundation.org
The Vitamin C Foundation
P.O. Box 73172
Houston, TX 77273
Toll Free: (800) 894-9025
Phone: (281) 443-3436

For Information and to Order Lypo-Spheric™ Vitamin C:

http://www.LivonLabs.com
LivOn Laboratories
2654 W. Horizon Ridge Parkway, Suite B-5, Dept. 108
Henderson, NV 89052
Phone: (702) 255-0265
Orders: (800) 334-9294
Customer Service: (866) 682-6193

For Information and to Order the Book, *Stop America's #1 Killer,* **by Dr. Thomas E. Levy:**

http://www.LivonBooks.com
Orders: (800) 334-9294

To Contact Owen Fonorow:

http://www.VitaminCFoundation.org/owen(contact form)
P.O. Box 3097
Lisle, IL 60532

To Contact Sally Snyder Jewell:

http://www.sallyjewell.com
Toll Free: (877) S-JEWELL (877-753-9355)

To Order Additional Copies of the Book, *Practicing Medicine Without a License? The Story of the Linus Pauling Therapy for Heart Disease,* **by Owen Fonorow and Sally Snyder Jewell, visit:**

http://www.lulu.com/paulingtherapy

Recommended Reading

Asimov I. 1972. Asimov's guide to science. New York. Basic Books Inc. 945 p.

Beisiegel U., Rath M., Reblin T., Wolf K., Niendorf A. 1990. Lipoprotein(a) in the arterial wall. Eur Heart J Aug;11 Suppl E:174-83.

Brecher A., Brecher H. 1998. Forty something forever, a consumer's guide to chelation therapy and other heart-savers, Healthsavers Press, 283 p.

Cameron E., Pauling L. 1993. Cancer and vitamin C. California. Camino Books. 278 p.

Cathcart R. 1981. Vitamin C, titrating to tolerance. Medical Hypotheses, 7:1359-1376. [online]. Available from: http://www.orthomed.com/titrate.htm. Accessed 2006 Jun 29.

Cheraskin E. 1984. The vitamin C connection. New York. Bantam Books. 336 p.

Cheraskin E. 1988. The vitamin C controversy: questions and answers. Wichita Kansas, Bio-Communications. 201 p.

Cheraskin E. 1993. Vitamin C who needs it. Alabama. Atticus Press & Company. 231 p.

Ellison S. 2006. The hidden truth about cholesterol-lowering drugs. [online]. Available from: http://www.health-fx.net/eBook.pdf. Accessed 2006 Jun 29.

Ely J. 2006. A physician's update on coenzyme Q10 in U.S. medicine. [online]. Available from: http://faculty.washington.edu/ely/coenzq10abs.html. Accessed 2006 Jun 29.

Fletcher A.E., Breeze E., Shetty P.S. 2003. Antioxidant vitamins and mortality in older persons: findings from the nutrition add-on study to the Medical Research Council Trial of Assessment and Management of Older People in the Community. Am J Clin Nutr Nov; 78(5):999-1010.

Furumoto K., Inoue E., Nagao N., Hiyama E., Miwa N. 1998. Age-dependent telomere shortening is slowed down by enrichment of intracellular vitamin C via suppression of oxidative stress. 1998: Life Sci.63(11):935-48.

Ginter E. 1982. Vitamin C in the control of hypercholesterolemia in man. Lipid Metabolism and Cancer. Ed.A. Hanck, Hans Huber. Bern.P. 135-152. [Ginter MEDLINE abstracts online] Available from: http://vitaminc foundation.org/forum/viewtopic.php?t=152. Accessed 2006 Jun 29.

Harwood H.J., Greene Y.J., Stacpoole P.W. 1986. Inhibition of human leukocyte 3-hydroxy-3-methylglutaryl coenzyme A reductase activity by ascorbic acid. An effect mediated by the free radical monodehydroascorbate. J. Biol. Chem., Vol. 261, Issue 16, 7127-7135, 06, [online]. Available from: http://www. jbc.org/cgi/content/abstract /261/16/7127. Accessed 2006 Jun 29.

Hickey S., Roberts H. 2004. Ascorbate the science of vitamin C. 246 p. [online]. Available from: http://www. lulu.com/content/55277. Accessed 2006 Jun 29.

Hickey S., Roberts H. 2004. Ridiculous dietary allowance. 151 p. [online]. Available from: http://www.lulu.com/content /92249. Accessed 2006 Jun 29.

Holford P. 1994. Vitamin C: how much is enough. [online]. Available from: http://www.vitamincfoundation.org/mega_1_1.html. Accessed 2006 Jun 29.

Huemer R. 1986. The roots of orthomolecular medicine: a tribute to Linus Pauling. New York. W.H. Freeman and Company. 290 p.

Huggins H., Levy T. 1999. Uniformed Consent: The hidden dangers in dental care. USA. Hampton Roads Publishing. 278 p.

Jacques, P.F., et al. 1997. Long-term vitamin C supplement use and prevalence of early age-related lens opacities. American Journal of Clinical Nutrition 66(October):911.

Klenner F. 1971. Observations on the dose and administration of ascorbic acid when employed beyond the range of a vitamin in human pathology. J of App Nutr Vol. 23, No's 3 & 4, Winter [online]. Available at http://www.orthomed.com/klenner.htm. Accessed 2006 Jun 29.

Levy T. Stop America's #1 killer. Las Vegas. Livon Books, 291 p.

Levy T. 2002. Curing the incurable: vitamin C, infectious diseases, and toxins. Las Vegas. Livon Books. 444 p.

Lewin S. 1976. Vitamin C: Its molecular biology and medical potential. London. Academic Press. 231 p.

Lopez D., Williams R., Miehlke M. 1994. Enzymes the fountain of life. South Carolina. Neville Press. 330 p.

Mares-Perlman J.A. Contribution of epidemiology to understanding relations of diet to age-related cataract. American Journal of Clinical Nutrition 66(October):739.

McDade L. 2002. Lowering Lp(a) with vitamin C, lysine and proline. [online]. Available from: http://vitaminc foundation.org/NCCAMgrant/. Accessed 2006 Jun 29.

Mindell E., Hopkins V. 1998. Prescription alternatives. Connecticut. Keats Publishing. 551 p.

Nathens A.B., Neff M.J., Jurkovich G.J., Klotz P., Farver K., Ruzinski J.T., Radella F., Garcia I., Maier R.V. 2002. Randomized, prospective trial of antioxidant supplementation in critically ill surgical patients. Ann Surg Dec;236(6):814-22.

Niendorf A., Dietel M., Beisiegel U., Arps H., Peters S., Wolf K., Rath M. 1990. Morphological detection and quantification of lipoprotein(a) deposition in atheromatous lesions of human aorta and coronary arteries. Anat Histopathol 1990; 417(2):105-11n R. [published erratum appears in Virchows Arch A Pathol Anat Histopathol 1991;418(1):86].

Osganian S.K., Stampfer M.J., Rimm E., Spiegelman D., Hu F.B., Manson J.E., Willett W.C. 2003. Vitamin C and risk of coronary heart disease in women. J Am Coll Cardiol. Jul 16; 42(2):253-5.

Padayatty S.J., Riordan H.D., Hewitt S.M., Katz A., Hoffer L.J., Levine M. 2006. Intravenously administered vitamin C as cancer therapy: three cases. CMAJ. Mar 28;174(7):956-7.

Pauling L. 1970. Vitamin C and the common cold. New York. Avon Book Company. 233 p.

Pauling L. 1986, 2006. How to live longer and feel better. Oregon University Press. 338 p.

Pauling L. 1992. A unified theory of cardiovascular disease. ION [video]. 60 min.

Linus Pauling in his own words. Edited by Barbara Marinachi, Simon & Schuster, 1995, ISBN 0-684-80749-1.

Rath M., Pauling L. 1990. Hypothesis: lipoprotein(a) is a surrogate for ascorbate, Proc Natl Acad Sci U S A. Aug;87(16):6204-7. [published erratum appears in 1991 Proc Natl Acad Sci U S A Dec 15;88(24):11588].

Rath M., Pauling L., 1990. Immunological evidence for the accumulation of lipoprotein(a) in the atherosclerotic lesion of the hypoascorbemic guinea pig. Proc Natl Acad Sci U S A 1990 Dec; 87(23):9388-90.

Rath M., Pauling L. 1991. A unified theory of human cardiovascular disease, J. Ortho Nutrition.

Reiter R., Robinson J. 1995. Melatonin. New York, Bantam Books. 399 p.

Sinatra S. 1999. Coenzyme Q10 and the heart. Chicago. Keats Publishing. 56 p.

Sommer A., West K. 1996. Vitamin A deficiency: health, survival, and vision. New York. Oxford Press. 438 p.

Stone I. 1972. The healing factor, vitamin C against disease [Also online. Available at: http://vitamincfoundation.org/ stone/] Accessed June 29.

Szent-Györgyi A. 1937. Oxidation, energy transfer, and vitamins. Nobel Lecture. [online]. Available from: http://nobelprize.org/nobel_prizes/medicine/laureates/1937 /szent-gyorgyi-lecture.pdf. Accessed 2006 Jun 29.

Williams R.J. 1977. The wonderful world within you. Wichita, Kansas. Bio-communications Press. 266 p.

Williams R.J. 1971. Nutrition against disease: environmental protection. Pitman Publishing Corp. 315 p.

Willis G.C. 1953. An experimental study of the intimal ground substance in atherosclerosis. Canad.M.A.J. Vol 69, p. 17-22. [online]. Available from: http://vitamincfoundation.org /pdfs/. Accessed 2006 Jun 29.

Willis G.C., Light A.W., Cow W.S. 1954. Serial arteriography in atherosclerosis. Canad.M.A.J. Dec 1954, Vol 71, p. 562-568. [online]. Available from: http://vitaminc foundation.org/ pdfs/. Accessed 2006 Jun 29.

Willis G.C., Fishman S. 1955. Ascorbic acid content of human arterial tissue. Canad.M.A.J., April 1, Vol 72, Pg 500-

503. [online]. Available from: http://vitamincfoundation.org /pdfs/. Accessed 2006 Jun 29.

Willis G.C. 1957. The reversibility of atherosclerosis. Canad.M.A.J., July 15, Vol 77., Pg 106-109. [online]. Available from: http://vitamincfoundation.org/pdfs/. Accessed 2006 Jun 29.

Wright J., Lenard L. 1999. Maximize your vitality and potency for men over 40. USA. Smart Publications. 256 p.

Yokoyama T., Chigusa D., Kokubo Y., et al. 2000. Serum vitamin C concentration was inversely associated with subsequent 20-year incidence of stroke in a Japanese rural community. Stroke.31:2287-2294.

Zandi P.P., Anthony J.C., Khachaturian A.S., Stone S.V., Gustafson Tschanz J.T., Norton M.C., Welsh-Bohmer K.A., Breitner J.C.S., for the Cache County Study Group. 2004. Reduced risk of Alzheimer disease in users of antioxidant vitamin supplements. Arch Neurol.61:82-88.

Selected References

Condado J.A., Waksman R., Gurdiel O., et al. Long-term angiographic and clinical outcome after percutaneous transluminal coronary angioplasty and intracoronary radiation therapy in humans. Circulation 1997;96:727-32.

Hampl J.S., Taylor C.A., Johnston C.S., Vitamin C Deficiency and Depletion in the United States: The Third National Health and Nutrition Examination Survey, 1988-1994, Am J Public Health 2004 May, 94(5):870-5.

Holford P, "Vitamin C: How Much is Enough?" MEGASCORBATE THERAPIES: Vitamin C in Medicine, Vol 1, 1, 1997.

Horie H., et al., "Association of acute reduction in Lp(a) with coronary artery restenosis after percutaneous transluminal angioplasty," Circulation, 1997 Jul 1;96(1):166-73, MEDLINE Cit ID: 97379632.

King S.B., Williams D.O., Chougule P., et al. Intracoronary beta radiation inhibits late lumen loss following balloon angioplasty: results of the BERT-1 trial [abstract]. Circulation 1997;96:1-219.

Klein, J., Williams D., Bonan R., Waksman R., Crocker I. The beta energy restenosis trial: update results and subgroup analysis.

Klezovitch O.; Scanu A.M.; Edelstein C., "Evidence that the fibrinogen binding domain of Apo(a) is outside the lysine binding site of kringle IV-10: a study involving naturally occurring lysine binding defective lipoprotein(a) phenotypes," J Clin Invest 1996 Jul 1;98(1):185-917.

Pazzucconi F., "Cholesterol synthesis inhibitors do not reduce Lp(a) levels in normocholesterolemic patients,"

Pharmacol Res 1996 Set-Oct;34(3-4);131-3 MEDLINE Cit ID: 972041326.

Phillips J., et al., "Lp(a): a potential biological marker for unruptured intracranial aneurysms," Neurosurgery 1997 May;40(5):1112-5; MEDLINE CIT ID: 97293308.

Price K.D.; Reynolds R.D.; Price C.S., "Hyperglycemia-induced latent scurvy and atherosclerosis: the scorbutic-metaplasia hypothesis,": Med Hypotheses 1996 Feb;46(2):119-29.

Rath M. und Niedzwiecki A. (1996), Nutritional Supplement Program Halts Progression of Early Coronary Atherosclerosis Documented by Ultrafast Computed Tomography. Journal of Applied Nutrition 48.

Stein J.H.; Rosenson R.S., "Lipoprotein Lp(a) excess and coronary heart disease," Arch of Intern Med 1997, Jun MEDLINE Cit ID: 97326406.

Stubs P., et al., "A prospective study of the role of Lp(a) in the pathogenesis of unstable angina," Eur Heart J 1997, Apr 18(4):603-7, Cit ID: 97276136.

Teirstein P.S., Massullo V., Jani S., et al. Catheter-based radiotherapy to inhibit restenosis after coronary stenting. N EnglJ Med 1997;336:1697-703.

Acknowledgments

It has been an honor to work with and learn from William Decker. We all owe a debt of gratitude to him and his company. Bill is a true entrepreneur in every sense of the word, having taken one of his companies public years ago. If you have been lucky enough to watch an American entrepreneur in action, then you know the magic. It was not an easy decision for Bill to form another company, Tower Orthomolecular Laboratories, having retired to the good life, though without his having taken that step this book would have never been written, for it is really a book of stories. Each case was an experiment, and none would have occurred had Bill lacked the knowledge, experience, perseverance, benevolence, and humanitarian instincts to face and overcome the obstacles and regulatory hurdles. Greed certainly wasn't the motivating factor, as no one gets rich selling vitamin C products.

Sally Jewell is an important part of this story in her own right and it has been my pleasure to work closely with her in the writing of this book. Not only has she contributed her formidable writing and editing skills to this project, but this book and a great many Pauling therapy users have benefited immeasurably from her thorough Pauling therapy knowledge and extensive experience, which has now spanned many years.

My thanks go to Jeff Fenlason, Carol Smith, William Cook, and the late Eli Raber for allowing us to invade their privacy. You are true heroes. Thanks go as well to the many others who have allowed the use of their names, such as Alan Oliver of Australia whose stories, for reasons of space, could not be included in this volume. Every story is unique and incredible, and additional case studies will be included in The Pauling Therapy Handbook, Volume II.

The contents of this book would not have been possible without the scientific contributions of pharmacology professors and authors Steve Hickey and Hilary Roberts. Their pursuit of the truth is an inspiration. We are grateful as well to author, medical doctor and attorney Thomas Levy for his scholarly persistence in pursuit of the facts and all that he has written and done for the furtherance of vitamin C education, as well as to Livon Labs for publishing Dr. Levy's latest work. We are also indebted to all of the earlier writers who wrote about vitamin C and from whom we have learned so much.

I would like to thank Steve Stone for allowing The Vitamin C Foundation to publish on-line without royalty fees the book written by his father, Irwin Stone, entitled, *The Healing Factor: "Vitamin C" Against Disease.*

I would also like to acknowledge and credit the late Jay Patrick, founder of the Alacer Corporation, for alerting us to the work of the Canadian doctor George C. Willis.

A special thanks to all participating members of the Vitamin C Foundation forum, who likely never realized that they were checking my ideas and sources and finding references by commenting on my posts. Interactive Internet forums are a powerful method of peer review and are akin to having a world-wide brain trust checking and challenging one's work. Even the smallest errors are rarely ignored.

This book contains many quotations and we have attempted to quote each accurately, especially those of Linus Pauling. The Pauling quotations are from his books, interviews, and the *Unified Theory* lecture videotaped by the British Institute of Optimum Nutrition.

I do not speak for Linus Pauling. I have not tried to lead or mislead anyone. All opinions, especially the unquoted negative opinions of prescription drugs, and any mistakes are my own.

I am grateful to my mother, Billie Valentine-Fonorow, for her work as my chief editor over the past dozen years and for encouraging me to put my material into book form. I thank you as well, Mom, for introducing me to vitamin C and Linus Pauling.

This book has been fun to write and is in part more readable because of Writer's Workbench, an artificial intelligence embodied in a computer program that was originally conceived and developed on the UNIX operating system by AT&T Bell Laboratories. Using Writer's Workbench is a lot like having HAL from the movie *2001* looking over your shoulder. The rights have been acquired and the software is now sold by EMO, Inc. for PCs running Windows and MS Word.

I am grateful to Linda Pauling-Kamb for providing the pictures of her father and for our conversations over many years.

Thanks go to my parents, who taught me the value of truth and honesty, and I am especially grateful for my dad's oft-uttered refrain, "Above all, think for yourself!" Thanks as well to my brother, Michael, without whom there would be no Vitamin C Foundation.

Lastly, I admit that this writing was inspired because of all the pharmaceutical companies that advertise their prescription drugs on American television. The rampant falsehoods in these inane commercials broadcast to millions could no longer be ignored.

Index

The Great Suppression

The Pauling Therapy Handbook

Volume II

Available soon

13744009R00163

Made in the USA
Lexington, KY
17 February 2012